Real Life Maths Skills

edited by Colin Ashbee
Head of Year,
Acland Burghley School, London

Heinemann Educational Books
London

Heinemann Educational Books Ltd
22 Bedford Square, London WC1B 3HH
LONDON   EDINBURGH   MELBOURNE   AUCKLAND
HONG KONG   SINGAPORE   KUALA LUMPUR   NEW DELHI
IBADAN   NAIROBI   JOHANNESBURG
PORTSMOUTH (NH)   KINGSTON

© Heinemann Educational Books Ltd 1984
First published by Scholastic Publications Ltd 1981
New edition 1984
Reprinted 1985, 1986

PUPILS BOOK   ISBN 0 435 10527 2
TEACHERS BOOK   ISBN 0 435 10528 0

NOTES

Metrication – In order to reflect real life in Britain today, both metric and imperial systems are used in this book. In some cases equivalents are given, and in others one system only is used in order to avoid confusion and complication. A reference section on using metric measures appears on p. 123, and a further note on p. 22.

Prices – All prices and charges quoted (e.g. the telephone charges on p. 24) were correct at the time of publication, but are obviously subject to change.

Acknowledgements

This book has been adapted from *Real Life Math Skills* originally published by Scholastic Inc, New York; and from *Real Life Maths* published by Ashton Scholastic, Australia, from which illustrative material has also been used.

The publishers would like to thank the following for permission to reproduce copyright material: .Abbey National Building Society for the form on pp. 44-5; Barclays Bank Limited for the material on pp. 31-37, 48, 49, and 52; British Telecom for extracts from the telephone charges leaflet on p. 24; the Controller of Her Majesty's Stationery Office and the Inland Revenue for sections of the income tax return form (Crown copyright) on p. 50; the Post Office for extracts from the postal rates leaflet on p. 26.

Printed and bound in Great Britain by Butler & Tanner Ltd, Frome and London

Who needs Real Life Maths?

To avoid Frank and Ernie's problems, read on . . .

You have been doing maths for years.

Now **Real Life Maths Skills** will give you experience in using maths in everyday life.

To get and keep a job, you need to use maths skills.
To run a home or a workshop, you need maths skills.
In sport, travel, shopping—you'll use maths every day.

In each **Unit** there are sections to help you learn and understand how to **use** your maths skills.

Most lessons have a **Fact Box**. In it is the information you might need to do the exercises.

You will be asked to calculate—to add, subtract, multiply and divide. If you need help to do this, go to the **Reference Section**. In the **Glossary**, you will find the meanings of words that you may need. Keep a working pad by you all the time. You will need it for calculations.

The **Looking Back** and **Skills Survey** pages at the end of each unit test your progress.

Contents

5

UNIT 1

A pocket calculator can save you a lot of time in solving or checking maths problems. Calculators are here to stay. They are getting better and cheaper all the time.
But a calculator is no use unless you understand the maths skills needed and can first do the calculations with pencil and paper. *This unit shows you how to get the best use from your calculator.*

How to Use a Calculator

Choosing a Calculator

There are hundreds of different calculators available. All pocket calculators add, subtract, divide and multiply. Some inexpensive calculators also have percentage (%), memory, and other keys. You may not need these keys for everyday use.

Before buying a calculator:
Read the instruction booklet.
Check whether the calculator does what you will need it to do.
If you plan to use it a lot, can you get an adaptor to plug it into a power point and save batteries?
Check that the window displays at least 8 digits.

Some hints for calculator owners:
Read the instruction booklet and try the examples given several times.
Check the accuracy:

1. Switch on

2. Press Clear [C]

3. Enter [1] [·] [1] [1] [1] [1] [1] [1] [1]

4. Press [×] [=] (The display should show 1.2345678) `1.2345678`

Keep your calculator away from the heat.
Keep a spare battery handy.
Always switch off after use.

Getting to Know your Calculator

For a calculator to be useful to you, you must tell it what you want it to do. When you use your calculator, first estimate (or guess) the answer to the problem. Then check your answer by doing the steps on the calculator.

This calculator is a common one.
The keys must be pressed in
the right order to get the
right answer.
**Read the instructions for
your calculator. It may be
different from this one.**

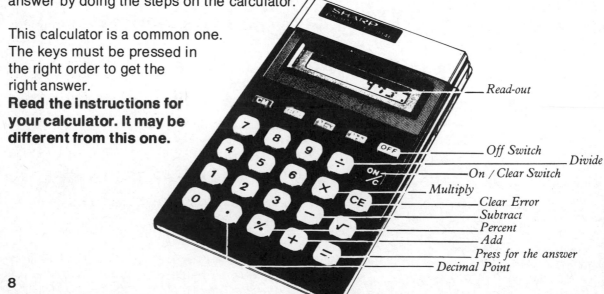

Read-out
Off Switch
Divide
On / Clear Switch
Multiply
Clear Error
Subtract
Percent
Add
Press for the answer
Decimal Point

Here is an example that shows you how to use your calculator.

To add: 12 + 35

a. Switch on.

b. Press [C] to clear the machine.

c. Press [1] and then [2] for 12.   Read out shows   `12.`

d. Press [+]   You want to add.

e. Press [3] and then [5] for 35.   Read-out shows   `35.`

f. Press [=] to get answer.   Read-out shows   `47.`

Now do this: 27 + 45 − 39
Press the keys in this order:

[C][2][7][+][4][5][−][3][9][=]   Read-out shows   `33.`

Circle the letter for the correct answer in Exs 1-4.

1. To find 35 + 8, which is the correct order for pressing the keys?

a. [C][3][5][+][8][=]

b. [3][5][C][+][8][=]

c. [C][3][5][8][+][=]

d. [3][5][+][C][=][8]

3. To find 7 × 8 + 4, which is the correct order?

a. [C][7][8][×][+][4][=]

b. [7][×][C][8][=][+][4]

c. [7][8][4][×][+][C][=]

d. [C][7][×][8][+][4][=]

2. To find 17 + 23 − 8, which is the correct order?

a. [1][7][+][C][2][3][8][−][=]

b. [1][+][7][2][3][−][C][=][8]

c. [C][1][7][+][2][3][−][8][=]

d. [C][1][7][2][3][+][−][8][=]

4. Choose the correct operation ([+], [−], [×], or [÷]).

a. [3][ ][5][=][8]

b. [4][ ][5][=][20]

c. [13][ ][6][=][7]

d. [18][ ][3][=][6]

9

5. Fill in the keys you must press to find the answer to each problem.

a. 31 + 23 　 [ ] [3] [1] [+] [2] [3] [=]

b. 17 ÷ 11 　 [C] [ ] [ ] [ ] [ ] [ ]

c. 49 ÷ 7 　 [ ] [ ] [ ] [ ] [ ]

d. 36 × 12 　 [ ] [ ] [ ] [ ] [ ] [ ]

e. 3 + 7 + 9 − 8 　 [ ] [ ] [ ] [ ] [ ] [ ] [ ] [ ]

f. 17 − 6 + 11 − 2 　 [ ] [ ] [ ] [ ] [ ] [ ] [ ] [ ] [ ] [ ]

Estimate, then Calculate

Some calculations are so easy that it is quicker to do them in your head than with your calculator. In real life, other calculations would take a long time. Remember, first you decide which operation (+, −, ×, ÷ or =) you need, then estimate the answer and press each key in the correct order.

1. Fill in the keys you must press to find the answer to each problem.

a. 1793 + 897 　 [C] [1] [7] [9] [3] [+] [8] [9] [7] [=]

b. 5462 − 3788 　 [ ] [ ] [ ] [ ] [ ] [ ] [ ] [ ] [ ] [ ]

c. 645 × 52 　 [C] [6] [4] [5] [×] [5] [2] [=]

d. 1934 × 17 　 [ ] [ ] [ ] [ ] [ ] [ ] [ ] [ ]

e. 487 ÷ 35 　 [C] [4] [8] [7] [÷] [3] [5] [=]

f. 2776 ÷ 120 　 [ ] [ ] [ ] [ ] [ ] [ ] [ ] [ ]

g. 16.04 + 19.72 　 [C] [1] [6] [•] [0] [4] [+] [1] [9] [•] [7] [2] [=]

h. 131.6 + 52.07 　 [ ] [ ] [ ] [ ] [ ] [ ] [ ] [ ] [ ] [ ] [ ] [ ]

i. 1.01 + 59.8 − 33.63

[ ] [ ] [ ] [ ] [ ] [ ] [ ] [ ] [ ] [ ] [ ] [ ] [ ] [ ] [ ] [ ] [ ]

j. 8.4 × 5.31 　 [C] [1] [6] [•] [2] [4] [÷] [7] [=]

k. 16.24 ÷ 7 　 [ ] [ ] [ ] [ ] [ ] [ ] [ ] [ ] [ ]

l. 86.6 ÷ 0.5 　 [ ] [ ] [ ] [ ] [ ] [ ] [ ] [ ]

2. Find the answers to these money problems in the same way. (You do not have to press a key for the pound sign. Round answers to 2 decimal places.)

a. £93.50 + £126.65

[C] [9] [3] [•] [5] [0] [+] [1] [2] [6] [•] [6] [5] [=]

b. £231.70 + £454.95

(empty keys)

c. £187.50 − £99.75

(empty keys)

d. £200.57 − £146.80

(empty keys)

e. £32.25 × 19

(empty keys)

f. £134.50 × 28

(empty keys)

g. £13.25 ÷ 5

(empty keys)

h. £166.00 ÷ 36

(empty keys)

Calculating Fractions

You can use your pocket calculator to solve problems involving fractions.

To find $\frac{2}{7}$ of 175 $\frac{2}{7}$ of $175 = \frac{2}{7} \times 175$ $= \frac{(2 \times 175)}{7}$ $= 2 \times 175 \div 7$

On your calculator you press these keys: [C] [2] [×] [1] [7] [5] [÷] [7] [=]

1. Fill in the keys you would press to solve these problems:

a. $\frac{7}{10}$ of 240   [C] [7] [×] [2] [4] [0] [÷] [1] [0] [=]

b. $\frac{3}{5}$ of 195

(empty keys)

c. $\frac{2}{3}$ of 186

(empty keys)

d. $\frac{9}{12}$ of 252

(empty keys)

e. $\frac{11}{17}$ of 2040

(empty keys)

f. $\frac{7}{100}$ of 350

(empty keys)

Calculating Commission with a Percentage Key

A commission of 5% on £90 = £90 × 5%

On your calculator press these keys. [C] [9] [0] [×] [5] [%]

Do not press [=] key.

4.5

Read this as £4.50.
Some calculators do not show the last zero.

1. Fill in the keys you would press to solve these problems:

a. A commission of 8% on £252   [C] [2] [5] [2] [×] [ ] [ ]

b. A commission of 11% on £550   [ ] [ ] [ ] [ ] [×] [1] [1] [ ]

c. A commission of 20% on £7300   [C] [ ] [ ] [ ] [ ] [ ] [ ] [ ] [ ]

d. A commission of 7½% on £600   [ ] [6] [0] [0] [ ] [7] [.] [5] [ ]

Calculating Mark-up and Discount

A mark-up or discount of 20% on £1300 = £1300 × 20%
On your calculator press these keys:

[C] [1] [3] [0] [0] [×] [2] [0] [%]

260.

A marked-up price of 20% on £1300 = £1300 + (£1300 × 20%)
On your calculator press these keys:

[C] [1] [3] [0] [0] [×] [2] [0] [%] [+] [=]

1560.
(1300 plus 20%)

A discounted price of 20% on £1300 = £1300 − (£1300 × 20%)
On your calculator press these keys:

[C] [1] [3] [0] [0] [×] [2] [0] [%] [−] [=]

1040.
(1300 minus 20%)

2. Fill in the keys you would press to find the selling prices for:

a. A mark-up of 20% on £2   [C] [2] [×] [2] [0] [%] [+] [=]

b. A mark-up of 12½%(12.5%) on £1070

[C] [ ] [ ] [ ] [ ] [×] [ ] [ ] [ ] [ ] [ ] [ ]

c. A discount of 5% on £16.80   [C] [1] [6] [.] [8] [0] [×] [ ] [ ] [=]

d. A discount of 15% on £29.60   [ ] [2] [9] [.] [6] [0] [ ] [ ] [ ] [ ] [ ]

e. A mark-up of 200% on £4.50   [ ] [ ] [ ] [ ] [ ] [2] [0] [0] [ ] [=]

f. A mark-up of 6% on £3500   [ ] [ ] [ ] [ ] [ ] [ ] [6] [%] [ ]

12

Calculators with Memory

Solving real-life problems often takes more than one step. That's when the calculator's memory comes in really handy. This lesson will show you how to use your calculator in solving real-life problems.

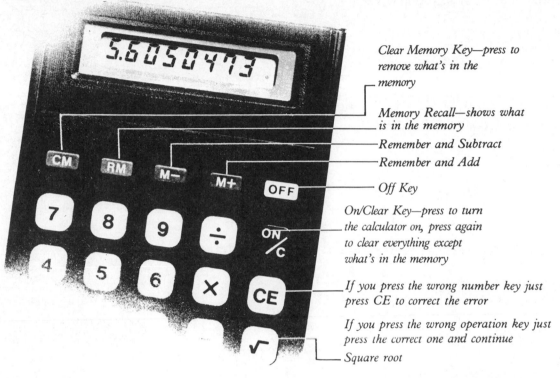

Clear Memory Key—press to remove what's in the memory

Memory Recall—shows what is in the memory

Remember and Subtract

Remember and Add

Off Key

On/Clear Key—press to turn the calculator on, press again to clear everything except what's in the memory

If you press the wrong number key just press CE to correct the error

If you press the wrong operation key just press the correct one and continue

Square root

The following examples show you how problems are solved with a calculator.
A £35 dress is on sale at 15% off. An £18 jumper is on sale at 10% off. How much would you pay for the two items? Press these calculator keys in order:

[C] [3] [5] [×] [1] [5] [%] [−] [M+]   Read-out: 29.75

[1] [8] [×] [1] [0] [%] [−] [M+]   16.20

[MR]   45.95

[CM]

1. Write the read-out for each step and find the total cost of these three items: a squash racket costs £16.40 and has a mark-up of 15%; a basketball costs £17.20 and has a mark-up of 25%; a set of darts costs £3.80 and has a mark-up of 20%.

Squash racket [C] [1] [6] [•] [4] [×] [1] [5] [%] [+] [M+] ....

Basketball   [1] [7] [•] [2] [×] [2] [5] [%] [+] [M+] ....

Darts   [3] [•] [8] [×] [2] [0] [%] [+] [M+] ....

                                    Total cost [MR] .... [CM]

2. Find the total cost. Round your answers to the nearest penny.

Shoes £45, 10% off

[C] [4] [5] [×] [1] [0] [%] [−] [M+] ....

Socks £1.55, 5% off

[1] [.] [5] [5] [×] [5] [%] [−] [M+] ....

Trousers £17.99, 15% off

[1] [7] [.] [9] [9] [×] [1] [5] [%] [−] [M+] ....

Shirt £10.50, 10% off

[1] [0] [.] [5] [0] [×] [1] [0] [%] [−] [M+] ....

Total [MR] ....

[CM]

3. Find the total selling price of these items.

Biscuits £0.50, mark-up 10%

[C] [.] [5] [0] [×] [1] [0] [%] [+] [M+] ....

Sugar £0.68, mark-up 12%

[.] [6] [8] [×] [1] [2] [%] [+] [M+] ....

Coffee £0.95, mark-up 25%

[.] [9] [5] [×] [2] [5] [%] [+] [M+] ....

Flour £0.32, mark-up 15%

[.] [3] [2] [×] [1] [5] [%] [+] [M+] ....

Total [MR] ....

[CM]

4. Fill in the keys you would press to solve these problems and find the total value:

TV Sets £250, discount 20%

[C] [2] [5] [0] [ ] [ ] [ ] [%] [−] [M+]

Radios £25, discount 35%

[2] [5] [×] [ ] [ ] [ ] [−] [ ]

Kettles £14.50, discount 10%

[ ] [ ] [ ] [ ] [ ] [×] [1] [0] [%] [ ] [M+]

Stereos £399, discount 5%

[3] [9] [9] [ ] [ ] [ ] [ ]

Total [M+] ....

[C]

14

UNIT 2

Do you end every week with the words, "Where has all the money gone?" If you answer "Yes", this unit is for you.

Your Daily Expenses

Paying for a Meal

Plan ahead . . . How much cash do you have? In most problems involving money, **addition** is the key. The prices of what you order should add up to the total amount you pay. This section provides practice in finding the total cost of a meal.

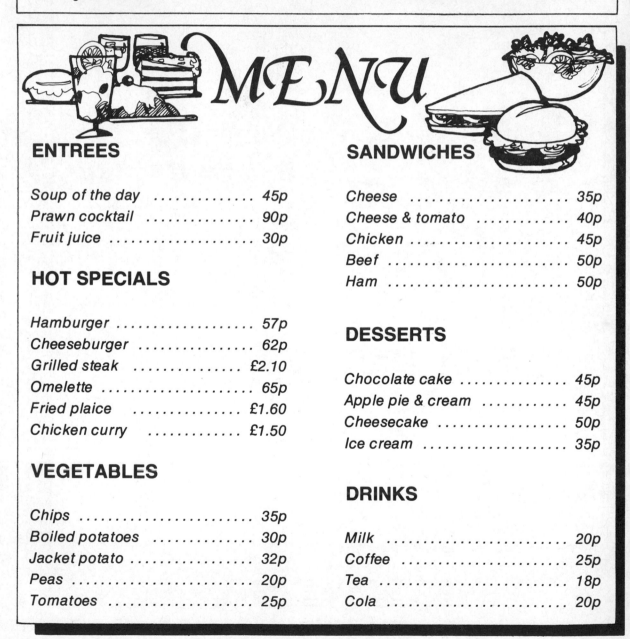

MENU

ENTREES

Soup of the day	45p
Prawn cocktail	90p
Fruit juice	30p

HOT SPECIALS

Hamburger	57p
Cheeseburger	62p
Grilled steak	£2.10
Omelette	65p
Fried plaice	£1.60
Chicken curry	£1.50

VEGETABLES

Chips	35p
Boiled potatoes	30p
Jacket potato	32p
Peas	20p
Tomatoes	25p

SANDWICHES

Cheese	35p
Cheese & tomato	40p
Chicken	45p
Beef	50p
Ham	50p

DESSERTS

Chocolate cake	45p
Apple pie & cream	45p
Cheesecake	50p
Ice cream	35p

DRINKS

Milk	20p
Coffee	25p
Tea	18p
Cola	20p

Look at the menu to find the price of each item. Write the prices and then add up to find the total cost of each meal. Ex. 1 is done for you.

1. Hamburger 57p
  Milk 20.p
  Total _____77p

2. Chicken sandwich ....
  Apple pie & cream ....
  Total _____

3. Ham sandwich ....
  Coffee ....
  Total _____

4. Cheeseburger ....
  Chips ....
  Cola ....
  Total _____

5. Cheese & tomato
  sandwich ....
  2 jacket potatoes ....
  Coffee ....
  Total _____

6. Chicken curry ....
  Ice cream ....
  Coffee ....
  Total _____

7. Cheese sandwich ....
  Tea ....
  Total _____

8. Soup ....
  Omelette ....
  Chips ....
  Tea ....
  Total _____

9. Fruit juice ....
  Fried plaice ....
  Chips ....
  Peas ....
  Cheesecake ....
  Total _____

10. Prawn cocktail ....
  Grilled steak ....
  Boiled potatoes ....
  Tomatoes ....
  Coffee ....
  Total _____

11. Soup ....
  2 jacket potatoes ....
  Milk ....
  Total _____

12. Hamburger ....
  Chips ....
  Chocolate cake ....
  Coffee ....
  Total _____

On Your Own

Michael works part-time taking orders in the café. On the order he uses abbreviations (short forms) of words, e.g. Chk Cry for Chicken Curry. Match his abbreviations with items on the menu and calculate the total cost for the following order:

ORDER		
Item	Price	Item names in full
1 Sp	0.45p	1 Soup of the day
2 Ch burg		
2 Bkd pot		
1 Choc cke		
1 Mlk		
1 Cffe		
	Total	

a. Now list the items that you would like to order. Calculate the total cost of your meal.

b. You find you have only £2 to spend on a meal. List the items you would like to order **and** can afford to pay for. (You do not have to spend all your money.)

Keeping Track of Expenses

How often do you say, "Where has all the money gone?" This section, which uses continuous subtraction, will help you keep track of your expenses.
Continuous subtraction means taking each new expense from the last balance.

Complete this week's calendar of expenses. Subtract the expenses, or amount paid, from the balance. Write the difference on the new balance line.

EXPENSE

	SUNDAY	MONDAY	TUESDAY	WEDNESDAY
Start-of-day balance	£30.00	29.10		
Expense	Sunday papers 0.55	Bus fares 0.40	Bus fares 0.40	Bus fares 0.40
New balance	29.45			
Expense	Pen 0.35	Snack 1.34	Birthday card 0.34	Squash court hire 2.10
New balance	29.10			
Expense		Stamps 0.48	Present 2.60	Drink at squash club 0.52
New balance				
Expense		Magazine 0.60	Book 1.50	Newspaper 0.18
New balance				
Expense			Light bulb 0.43	Toothpaste 0.42
New balance				
Expense			Chocolate 0.22	
New balance				
Expense			Fuse wire 0.27	
New balance				
Expense				
End-of-day balance	29.10			

Fact box

CALENDAR

THURSDAY		FRIDAY		SATURDAY
Bus fares	0.40	Bus fares	0.40	
Postal order	0.45	Snack	1.32	
Dry cleaning	2.60	Record	2.45	
		Disco ticket	0.75	
		Drink	0.82	

On Your Own

On Saturday morning John buys a newspaper. In the afternoon he travels by bus to watch his local football team play. At the match he buys a programme and has a hot-dog at half-time. In the evening he walks to the cinema. After the film John buys some fish and chips for his supper. Keep a track of John's expenses on Saturday, writing in the amount of money you think he spends. Calculate how much he has left at the end the week.

How to Save on Public Transport

Single fare? Day return? Monthly ticket? Weekly rate? Which is the best buy? There are often many different rates for the same journey, depending upon time of travel and number of journeys made. This section will show you that it pays to work out your travelling carefully.

This chart shows the various fare rates from Central Newtown to three different destinations.

MID COUNTIES BUS COMPANY					
	DESTINATION	ORDINARY SINGLE	WEEKLY	MONTHLY	CHEAP DAY RETURN
Central Newtown to	Oxley	65p	£6.00	£20.00	80p
	Belton	45p	£4.20	£19.20	60p
	Cranston	25p	£3.50	£12.00	40p
Cheap day returns can NOT be used between 8–10 a.m. and 4–6 p.m.					

1. Tom Spencer lives in Oxley and travels to and from work in Central Newtown five days each week.

a. What does an ordinary single ticket cost? ....65p ◄——— from the chart under Ordinary Single

b. How many single trips does he make in one week? ....10 ◄———
2 trips a day
× 5 days a week
10 trips a week

c. How much is a weekly ticket? £6.00

d. What is the cost of each of his trips on the weekly fare plan? ....60p ◄——— £6.00 ÷ 10 = 60p

e. How many single trips does he make in one month (4 weeks)? ....40 ◄———
10 trips a week
× 4 weeks a month
40 trips a month

f. How much is a monthly ticket? £20.00

g. What is the cost of each trip on the monthly fare plan? ....50p ◄———
    .50
40)20.00

h. Which fare plan is cheapest for Tom? monthly

2. Pat goes to Central Newtown and back home to Belton by bus six times a week for a month.

a. What does the ordinary single ticket cost? ....

b. How many single trips does Pat make in a week? ....

c. How much does a weekly ticket cost? ....

d. What is the cost of each ticket on the weekly fare plan? ....

e. How many single trips are made in a month? ....

f. How much is a monthly ticket? ....

g. What is the cost of each trip on the monthly fare plan? ....

h. Which fare plan is cheapest? ....

3. Next month, as well as her normal bus travel, Pat has to make ten extra trips by train at 50p each.

a. What is the cost of bus travel for the month if she buys four weekly tickets? ....

b. What is the cost of her train travel for the month? ....

c. What is the total cost of travel for the month? ....

d. Should Pat buy a monthly ticket for that month? ....

e. How much would she save by buying a monthly ticket? ....

4. Steve makes fourteen single trips a week between Cranston and Central Newtown.

a. How much is the ordinary single ticket? ....

b. How much will he pay each month if he buys ordinary single tickets? ....

c. How much is a monthly ticket? ....

d. How much will Steve save if he buys the monthly ticket? ....

5. Peter and Sue Morris both work part of the week in Newtown. They live in Oxley and travel to and from work four times a week. Sue's job allows her to travel at off-peak times (after 10 a.m. and before 4 p.m.) but Peter's does not. Find the cheapest fare plans for Peter and Sue for a month's travel. How much more than Sue does Peter have to pay for the month?

On Your Own

Make enquiries about special fare plans for public transport in your area. Pick a place where you could go to work and decide which fare plan would be best for you.

Ordinary single fare ....

Monthly rate ....

Number of trips you might make in a month ....

Cost of each single trip ....

Weekly rate ....

Number of trips you might make in a week ....

Cost of each single trip ....

Which fare plan is cheapest? ....

At the Supermarket

"Do I have to buy the whole thing?" What if you only want to buy half a cucumber, or three-quarters of a pound of steak? To calculate the cost of anything less than the whole item you have to use fractions.

First fill in the missing information in the meat section of the advertisement. Then use the advertised prices to calculate the total cost of each shopping list. Often you must calculate the fractional cost of an item. Where answers include fractions of a penny, calculate them to the next whole penny.

A fraction is part of a whole.

To find a fractional cost: ¾ of £3.72 = ?
1. Multiply the cost of the whole item by the numerator of the fraction:

$$3 \times 3.72 = 11.16$$

2. Divide the result by the denominator.

$$11.16 \div 4 = 2.79$$

numerator → $\dfrac{3 \times £3.72}{4} = \dfrac{3 \times £3.72}{4}$ ← denominator

$$= \frac{£11.16}{4}$$

Three-quarters of £3.72 = £2.79

1000 g (grams)	= 1 kg (kilogram)
750 g	= ¾ kg
500 g	= ½ kg
250 g	= ¼ kg

16 oz (ounces)	= 1 lb (pound)
12 oz	= ¾ lb
8 oz	= ½ lb
4 oz	= ¼ lb

1 whole   ½   ¼   ¾

In Britain most **pre-packed** items are calculated in grams.
Fresh foods are usually weighed in pounds and ounces.

Supersave Stores — The cut-price food stores!

FRUIT & VEG.

Potatoes	14p	lb
Cox's apples	30p	lb
Tomatoes	36p	lb
Cucumbers	32p	ea
Oranges	5 for 60p	

FRESH FISH

Cod fillets	£1.40	lb
Plaice	£1.80	lb
Haddock	£1.55	lb

BEVERAGES & CEREALS

Tea	£3.44	kg bag
Coffee	£4.80	750g tin
Cornflakes	52p	500g pkt.
Porridge oats	53p	750g pkt.

MEAT

Beef – topside	£1.84	lb
Normal price	£2.10	lb
Save	£	lb
Rump steak	£2.48	lb
Normal price	£2.80	lb
Save	£	lb
Lamb chops	£1.32	lb
Normal price	£1.50	lb
Save	£	lb
Pork sausages	68p	lb
Chicken pieces	£1.30	lb
Ham	£2.00	lb
Salami	£1.96	lb

1. 3 lb potatoes   £ 0·42 ← $3 \times 14 = £0.42$ (or 42p)

    ¾ lb pork sausages   £ 0·51 ← ¾ of $68 = \dfrac{3 \times 68}{4} = \dfrac{204}{4} = £0.51$

    1 kg bag tea   £ 3·44 ← from the ad.

                Total   £ 4·37

2. 2 lb tomatoes   ....

    1 lb rump steak   ....

    ½ lb plaice   ....

                Total   _____

3. 10 oranges   ....

    1 500g pkt cornflakes   ....

    ¼ lb ham   ....

                Total   _____

4. ½ cucumber   ....

    3 lb Cox's apples   ....

    ¾ lb cod fillets   ....

                Total   _____

5. 2½ lb chicken pieces   ....

    750g tin coffee   ....

    1 cucumber   ....

    ¼ salami   ....

                Total   _____

6. 5 oranges   ....

    1½ lb lamb chops   ....

    1 750g pkt porridge oats   ....

    2 lb haddock   ....

                Total   _____

7. 3 lb beef topside   ....

    4 lb tomatoes   ....

    1½ lb cod fillets   ....

    4 oz ham   ....

    1¼ lb pork sausages   ....

                Total   _____

On Your Own

From a supermarket advertisement in a newspaper, list the items you might buy on a weekly shopping trip. Calculate the total cost of your list.

.... £...............

....

....

....

....

                Total   _____

Long-distance Rates

Is telephoning home by STD really the next best thing to being there? In this section you can practise using a chart to calculate the cost of telephone calls. See how to save money by phoning at different times of the day and on different days of the week. It's your choice: dialled direct, personal call, transferred charge, operator-connected. Maybe you'll decide that being there is the only thing!

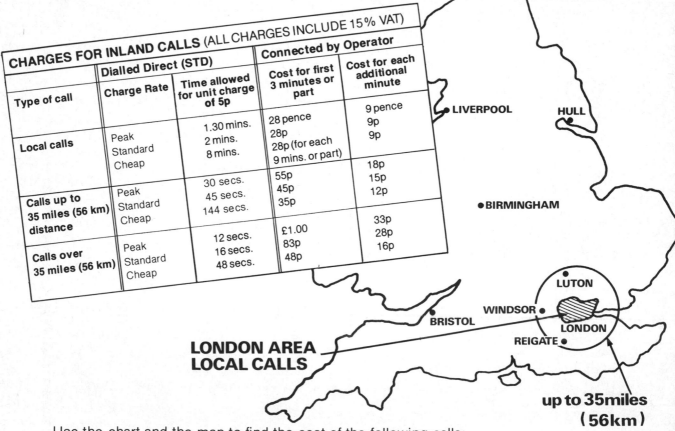

CHARGES FOR INLAND CALLS (ALL CHARGES INCLUDE 15% VAT)

Type of call	Dialled Direct (STD)		Connected by Operator	
	Charge Rate	Time allowed for unit charge of 5p	Cost for first 3 minutes or part	Cost for each additional minute
Local calls	Peak	1.30 mins.	28 pence	9 pence
	Standard	2 mins.	28p	9p
	Cheap	8 mins.	28p (for each 9 mins. or part)	9p
Calls up to 35 miles (56 km) distance	Peak	30 secs.	55p	18p
	Standard	45 secs.	45p	15p
	Cheap	144 secs.	35p	12p
Calls over 35 miles (56 km)	Peak	12 secs.	£1.00	33p
	Standard	16 secs.	83p	28p
	Cheap	48 secs.	48p	16p

LIVERPOOL HULL

● BIRMINGHAM

LUTON

WINDSOR ●

BRISTOL LONDON

REIGATE ●

LONDON AREA LOCAL CALLS

up to 35 miles (56 km)

Use the chart and the map to find the cost of the following calls. Exs 1 and 2 are done for you. All calls are made from London.

1. To Bristol on Tuesday at 7 p.m., STD for 9 mins:
   No. of seconds allowed for
   each unit charge ..... *48.* ◄——— From the chart under STD (cheap rate)
   Length of call in seconds ..*540.* ◄——— 9 × 60 ◄— (60 secs = 1 minute)
   No. of unit charges ......*12.* ◄——— 540 ÷ 48 ◄— 11.25 (rounded up to 12)
        Total cost   *60p* ◄——— 12 × 5p

2. Call to Reigate, person to person, operator connected, on Sunday for 17 minutes.
   Rate for first 3 mins. .....*35p* ◄——— from chart, connected by operator (cheap rate)
   No. of additional minutes .....*14*...
   Rate for each additional min. ......*12p.*
        0.35 ◄——— 3-min. charge
        1.68 ◄——— rate × no. of additional mins. (12 × 14)
   Call fee   *£2.03*
   Surcharge ....*51p* ◄——— personal call fee 51p (see FACT BOX)
        Total cost *£2.54* ◄——— call fee + surcharge (£2.03 + 51p)

3. Call to Luton on Monday at 10 a.m., STD
   for 6 minutes:
   No. of secs. for each unit ch. ....
   Length of call in secs. ....
   No. of unit charges ....
        Total cost ———

4. To Hull, operator connected, on Friday at
   5 p.m. for 8 mins.
   Rate for first 3-min. period ....
   No. of additional minutes ....
   Rate for each additional min. ....
   Call fee ....
        Total cost ———

5. Local call, 3 p.m., for 12 mins. on
   Monday, STD.
   No. of minutes for each
   unit ch. ....
   Length of call in minutes ....
   No. of unit charges ....
        Total cost ———

6. Transferred charge call to Liverpool, on
   Tuesday at 12 a.m. for 5 minutes
   (transferred calls must go through the
   operator).
   Rate for first 3-min. period ....
   No. of additional minutes ....
   Rate for each addl. min. ....
   Call fee ....
   Surcharge ....
        Total cost ———

7. How much money is saved if the call to
   Liverpool in Ex. 6 is made as an STD call
   instead of a transferred charge call?
   ....

8. What is the difference in cost between
   the following two calls?
   a. To Windsor on Wednesday at 8 p.m.
   STD for 10 minutes:
   No. of secs. for each unit ch. ....
   Length of call in secs. ....
   No. of unit charges ....
        Total cost ———

   b. To Windsor, operator connected, on
   Wednesday at 8 p.m. for 10 minutes:
   Rate for first 3-min. period ....
   No. of additional minutes ....
   Rate for each additional min. ....
   Call fee ....
        Total cost ———

   The difference in cost
   between **a.** and **b.** is   ....

In the Post Office

Do you like to receive letters? That's the easy bit. Wait until you see how much mail costs to send.

Many post offices provide scales so that you can weigh your own mail and calculate the cost. This section will give you practice in reading scales and calculating the cost of sending letters and parcels by mail.

Rates for letters within the UK

Weight not over	1st Class	2nd Class
60 g	16p	12½p
100 g	23p	17p
150 g	29p	21p
200 g	36p	27p
250 g	43p	33p
300 g	50p	39p
350 g	58p	45p
400 g	66p	51p
450 g	74p	57p
500 g	82p	63p
750 g	£1.22	95p max
1000 g	£1.62	

Rates for parcels within the UK

Weight not over	National Rate	Local Rate
1 kg	£1.30	£1.10
2 kg	£1.67	£1.47
3 kg	£2.00	£1.80
4 kg	£2.20	£2.00
5 kg	£2.35	£2.15
6 kg	£2.50	£2.30
7 kg	£2.65	£2.45
8 kg	£2.80	£2.60
9 kg	£2.95	£2.75
10 kg	£3.10	£2.90

Postal Rates Inland

Royal Mail

1. Letter Post

Write the weights shown on the scales for the items **a** to **g**. Find the postage costs using the information from the rate table.

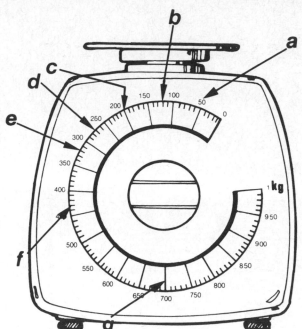

Article	Class	Weight	Cost
a	1st	50 g	16p
b	2nd		
c	2nd		
d	1st		
e	1st		
f	2nd		
g	1st		

2. Parcel Post

Calculate the mailing costs for parcels **h** to **n**. Draw arrows to the markings on the scale to show the weight of each parcel, and label each arrow-line. Article **h** is done for you.

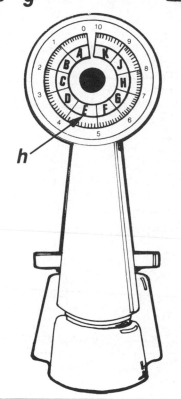

Article	Rate	Weight	Cost
h	local	4.5 kg	£2.15
i	national	2.4 kg	
j	national	750 g	
k	local	3.2 kg	
l	local	8.0 kg	
m	national	6.8 kg	
n	national	9.5 kg	

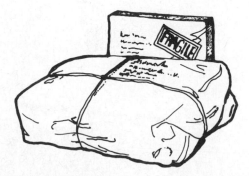

On your Own

★ Find out what you can about: Registered Letters, Recorded Delivery and the Cash on Delivery service.
★ What does Post Office Preferred (POP) mean?

Looking Back

1. Write a bill for one dinner that includes the following: tomato soup—£0.86; steak and salad—£3.65; blackcurrant cheesecake—£0.94; coffee—£0.36. Calculate the total cost.

CAFÉ

Item	Price

Thank You—Call again!   Total

3. Fill out a sales docket for: 3 pens—£0.39 each; 1 ruled pad—£0.50 each; 2 notebooks—£0.65 each; 4 pencils—10 for £1.20. Calculate the total cost.

PAD & PENCIL SUPPLY CO.

ItemDescription	UnitPrice	Qty	Total Cost
		Total	

2. Your cash balance on Monday morning was £50.00. Your expenses from Monday to Friday were: £12.60, £4.52, £9.26, £6.05 and £11.54. How much did you have at the end of each day? What was your end-of-week balance?

Balance
Monday expenses
New balance
Tuesday expenses
New balance
Wednesday expenses
New balance
Thursday expenses
New balance
Friday expenses
New balance
End-of-week balance

4. A monthly ticket which is good for sixty trips costs £71.70. A weekly ticket good for fourteen trips costs £20.80. The single ticket fare is £3.25.
Which ticket should the following people buy?

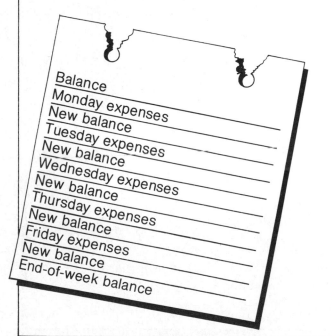

a. Jane Wood who makes twelve trips each week:

....

b. Bill Russell who goes to work and returns home three times a week for a month:

....

5. A sack of potatoes costs £6.72. Write the cost of the following fractional parts of the sack:

$\frac{3}{4}$ ....

$\frac{2}{3}$ ....

$\frac{1}{2}$ ....

$\frac{1}{3}$ ....

$\frac{1}{4}$ ....

Potatoes
King Edward
£6.72 a sack

6. What is the difference in cost between the following two calls?
a. To Birmingham, STD, on Monday at 5 p.m., for 5 minutes. You are allowed 16 seconds for each unit charge (5 pence).
No. of secs. for each unit charge ....
Length of call in seconds ....
No. of unit charges ....
       Total cost ....

b. To Birmingham, operator connected, on Monday at 5 p.m., for 5 minutes.
The cost for each 3-minute period is 83p
Each additional minute costs 28p.
Rate per 3 minutes ....
No. of 3 min. periods charged ....
No. of additional mins. ....
Rate for each additional min. ....
       Total cost ....

The difference in cost between **a.** and **b.** is ....

7. Write the weight shown on the scale for each parcel. Calculate the postage costs at the following national rates:

Weight not over	Cost	Article	Weight	Cost
1 kg	£1.30	a		
2 kg	£1.67	b		
3 kg	£2.00	c		
4 kg	£2.20	d		
5 kg	£2.35	e		

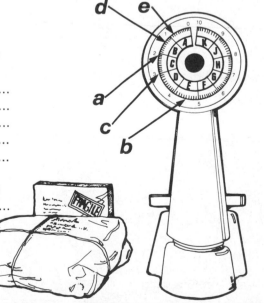

Skills Survey

You have seen how useful maths skills are in your daily activities. The exercises in this section will help sharpen your skills.

1. Add:

```
        10              457
         5             3214
    + 204               62
                     + 135
```

```
    £29.75           £10.00
      6.82           131.16
      0.49             0.08
    + 2.63          + 40.05
```

£4.55 + £0.89 + £24.50 = ...

2. Subtract:

```
    6879              341
  - 2765            - 265
```

```
    £8.25           £25.43
  - 4.15            - 9.39
```

£128.78 − £32.69 = ...

3. Multiply:

```
    2743             2135
  ×   50           ×   32
```

```
    0.87            65.23
  ×    3          ×  0.05
```

4.35 × 0.25 = ...

4. Divide:

```
    2  848          32  3968

   26  55.90        13  19.50
```

164.30 ÷ 62 = ...

5. Round each answer to the nearest penny:

```
      £5.14             £7.32
    × 0.03           ×  0.06
```

```
    5  £61.32        32  £73.40
```

£101.60 ÷ 48 = ...

6. $\frac{2}{3}$ of £9.72 = ...

$\frac{3}{4}$ of £8.56 = ...

$\frac{1}{2}$ of £253.64 = ...

$\frac{1}{4}$ of £672.84 = ...

Branching Off

Find out how a taxi meter works. Ask a local taxi driver how much the first distance costs and how much each additional kilometre or mile costs. Calculate the total costs of distances you might want to travel. Check for any extra charges that can be made.

UNIT 3

When you plan your expenses, take care of what you really need first. You might even save enough for that trip or car you've always dreamed about!

Managing Your Money

Cheque Account

What piece of paper becomes money with a few strokes of the pen? You may write one to pay a bill, or to buy something when you don't have enough cash with you. It's a cheque of course. In this section you will learn the basic steps in using a cheque account.

A cheque account lets you deposit money in the bank and take it out again by writing a cheque.

a. To fill in a deposit slip, follow the numbered steps in the example:

1. Write the date.
2. Name of bank and branch.
3. Print your name and account number.
4. Separate notes into denominations and write the amounts on notes line.
5. Separate coins into denominations and write amounts on coin lines.
6. List total amount of any cheques deposited on the cheques line.
7. Add the notes, coins and cheques and write this sum in the space provided.
8. Sign your name.

b. To write a cheque, follow the numbered steps in the example:

1. Write the date.
2. Write the name of the person or company to be paid.
3. Write the pounds amount in words e.g. fourteen pounds, and the pence amount in numerals e.g. 58 pence.
4. Write the amount of the cheque again, this time in numerals e.g. £14.58.
5. Sign the cheque.

Drawer: The person or company who issued the cheque.
Bank: The bank name on the cheque.
Branch: The name of the town and the address of the bank.

1. You have a £20 note, four £5 notes, and £2.85 in silver coins. You want to deposit these as well as the cheque from Ex. **b.** made payable to you.

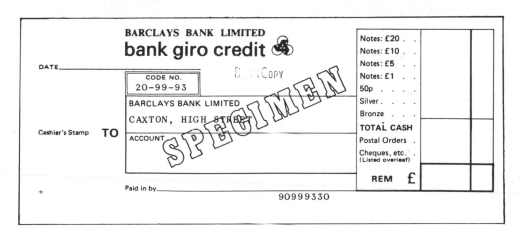

2. You have three £10 notes, six £5 notes, twenty £1 notes, £5 in silver and £4.72 in bronze coins. You want to deposit cheques for £7.50, £78.25, £10.00 and £21.87.

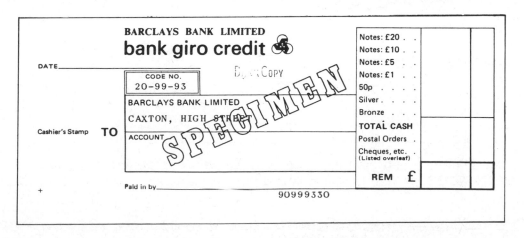

Write cheques for the payments in Ex. 3 and Ex. 4.

3. On 5 September, you bought a car radio from Dynamo Auto Supplies for £42.95.

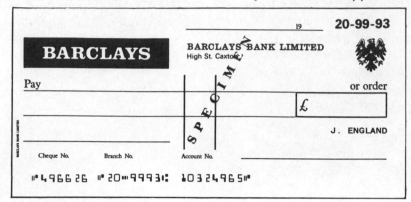

4. You cashed a cheque for £25.50 on 10 November. (Write **cash** on the line marked **Pay**.)

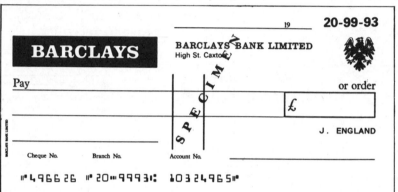

On Your Own

Find an item you would like to buy from a display advertisement in a newspaper. Write a cheque to pay for it.

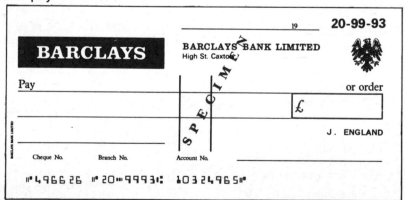

Find out the difference between "**crossed**" and "**open**" (uncrossed) cheques.
    REMEMBER: A cheque is worthless if there is no money in your account to cover it.

Balancing your Cheque Book

What piece of paper can "bounce"? A cheque can! If you write a cheque for more than the amount that you have in your account, your cheque will "bounce". That means the cheque is worthless and will be returned to you by the bank. You will still have to pay the amount you owe, plus an additional amount to the bank as a penalty charge. This section will help you to keep track of the money in your account.

There are two types of commonly used cheque books.

Counterfoil cheque book

A counterfoil (stub) remains each time a cheque is removed from the cheque book. The cheque stub has space to record details for each cheque as follows:

1. Date.
2. The name of the person or company paid.
3. What the cheque was for.
4. Old balance—the amount of money left in your cheque account after the last cheque was written.
5. Deposits—money deposited in your cheque account since the last cheque.
6. Total—amount in cheque account after adding any deposits.
7. This cheque—the amount written on this cheque.
8. New balance—final balance after subtracting this cheque.

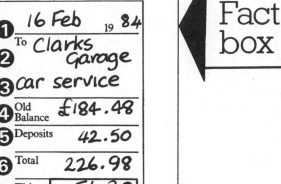

Fact box

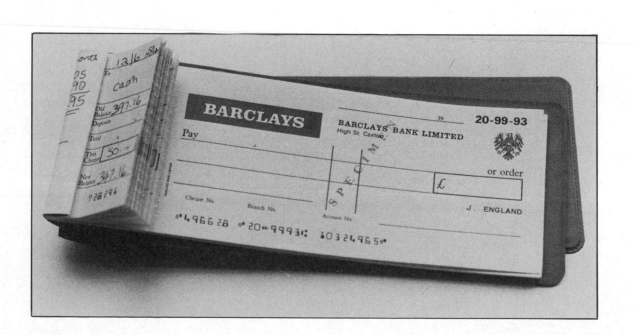

Statement cheque book

A statement of cheques and deposits is kept in a special section at the front of the cheque book. There are no cheque stubs. To record cheque details on the statement page, follow the numbered steps in the example:

1. Write the date.
2. Write the cheque number.
3. Write the name of the person or company you paid.
4. Write what the cheque is for.
5. Write the amount of the cheque.
6. Subtract the amount of the cheque from the old balance to find the new balance. To record a "Deposit" (credit) follow the numbered steps.
7. Write the date of the deposit.
8. Write "Deposit" and the description of it, if needed.
9. Write the total amount you have been credited with.
10. Add the credit to the old balance to find the new balance.

Date	Cheque No.	Cheques issued to/Credits paid in	Amount of cheque		Amount of credit		Balance	
		Balance Brought Forward					186	42
①6/2/84	②622	③Holts Sportswear ④Track suit	⑤16	99			⑥169	43
⑦15/2/84		⑧Deposit – Tax Refund			⑨23	62	193	05 ⑩
							Balance carried forward	

Note: The bank sends you a regular statement showing all deposits, cheques drawn and fees charged. You can use this to check your own entries.

1. Fill in the following cheque stubs with the payments and credits listed below. Starting balance 1 Sept £150.00:

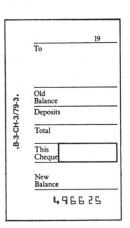
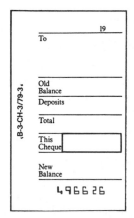

Chq. no.

624	3 Sept	Post Office (television licence)	£34.00
625	5 Sept	Cellar Restaurant (birthday dinner)	£16.50
	10 Sept	Credit (salary cheque)	£465.62
626	12 Sept	Cash (spending money)	£50.00
	12 Sept	Credit (tuition fee)	£42.00
627	14 Sept	Sun Tours (holiday deposit)	£94.00

36

2. Fill out the following cheque statement for the payments and credits listed below. Starting balance 1 Feb £350.00:

Date	Cheque No.	Cheques issued to/Credits paid in	Amount of cheque		Amount of credit		Balance	
		Balance Brought Forward					350	00
						Balance carried forward		

Chq. no.

532	1 Feb	Ashcroft Estates (rent)	£225.00
	2 Feb	Credit (salary cheque)	£594.24
533	5 Feb	North Thames Gas (gas bill)	£56.22
534	9 Feb	Green Hill Bookshop (gift)	£8.50
535	16 Feb	Cash (spending money)	£50.00

On Your Own

Record the deposit and cheques you might make the first month you are " on your own". Ask your teachers, family and friends which type of cheque book they prefer.

SURVEY

Number using counterfoil cheque books ....

Number using statement cheque books ....

Ask your bank which kind of cheque book is more popular ....

Does this agree with the results of your survey? ....

Discuss the advantages of each type of cheque book and decide which you prefer.

Find out what other services your bank can offer.

Savings

Saving for a holiday? New clothes perhaps? Regular deposits of money in a building society account are one way to save. It's safe and the building society pays you interest for the use of your money. In this section you will learn to use a building society account and see how interest in calculated.

<table>
<tr><td>Fact box ▶</td><td>

A **Passbook** is a record of your deposits, withdrawals, and interest earned.
Deposits are added to the balance.
Withdrawals are subtracted from the balance.
Interest is paid by the building society for the use of your money.
Interest = **Balance** (£) × **Rate** (%) × **Time** (years)
Example: Find the interest on £100 for 3 months at 9% yearly.
Interest = £100 × 9% × ¼ year ← 3 months = ¼ year
        = (100 × 0.09) × ¼ ← 9% = 0.09
        = 9 × 0.25 ← ¼ = 0.25
        = 2.25
Interest equals £2.25

</td></tr>
</table>

The interest **rate** can change. It often depends on the national minimum lending rate. Building societies use computers to calculate your interest and add it to your account regularly.

To put money into your building society account you fill out a **credit** slip and present it with your passbook to the cashier.

ACCORD BUILDING SOCIETY

Name ...Mr. John Peter.................................

............LOMAX..

Roll no. 2/   13756 543 – 9

Branch   Caxton

Date	Cashier's initials	Details	Withdrawals	Receipts	Balance
					468.54
7/9/83	C.A.	Fifty pounds ch		50.00	518.54
4/1/84	J.B.	Interest		26.32	544.86
13/1/84	C.A.	Sixty five pounds csh.	65.00		479.86

Read the facts carefully and answer the following questions.

1. Suppose you open an account on 1 January with a deposit of £200. How much interest will this money earn at the end of six months at an interest rate of 4% yearly? ....
What is your new balance if this interest is added to your account? ....

2. You want to withdraw £30.50 from your account on 1 July. Fill out this withdrawal slip. Use 10824 354 - 9 as your account number.

WITHDRAWAL NOTICE				**ABBEY NATIONAL** BUILDING SOCIETY
PLEASE USE BLOCK CAPITALS				

IMPORTANT: The Passbook must accompany this notice. Evidence of identity may be required. If the amount you require exceeds that normally repayable by the cashier, a cheque will be sent to your address. Cheques made payable to third parties cannot be cancelled or stopped by the Society.

ACCOUNT No.	Surname	Initials	Date / /

Address		Withdrawal Amount	**Cash**
Postcode			**Cheque**
Cheque payable to/ Special Instructions			**Total £**
		Investors Signature(s)	

OFFICE USE ONLY		
BALANCE OF ACCOUNT	£..............	
PRINCIPAL WITHDRAWN	£..............	
6 MONTHS INTEREST £..............	Branch No.	Cashier No. Cheque Number
INTEREST DEBIT £..............		
INTEREST PAID £..............	Chq. Signed Sig. Checked Certified	Identity Type Chq. Date
TOTAL WITHDRAWN £		

242/i/W/1

3. Fill in the missing entries in this passbook.

Name Winston Clive				**ACCORD** BUILDING SOCIETY	
HUGHES			Roll no. 2/ 10824 354 – 9		
			Branch High St. Newtown		
Date	Cashier's initials	Details	Withdrawals	Receipts	Balance
		Brought Forward			400.00
2/1/84	R.T.	Forty pounds ch.		40.00	....
14/1/84	B.H.	Eighty pounds ch.		....	....
22/1/84	B.H.	One hundred pounds csh.	....		....
12/2/84	R.T.	ch.		168.50	....
3/3/84	R.T.	Interest		19.25	....
25/3/84	R.T.	Fifty pounds csh.	50.00		....
31/3/84	K.M.	csh.	35.00		....
2/4/84	B.H.	Two hundred & ten pounds ch.		210.00	....

On Your Own

Find out what you can about:
Joint Accounts
"Save-As-You-Earn" schemes
The National Savings Bank
"Granny" Bonds

4. Higher interest rates can apply if you deposit your money for a fixed period of time.
If the money is withdrawn before the agreed time, you receive lower interest.
The Accord Building Society offers these yearly rates on time savings: 8% if the deposit is kept in the building society for 1 year, 9% on 2 year time deposit, and 10% on three year accounts. How much interest will the following amounts earn at the end of the first year?

Interest on			
Amount of deposit	1 year account	2 year account	3 year account
£50			
£75			
£135			

Budgeting

Earning money may be hard; spending it is very easy. That's why it is important to have a budget. When you plan your expenses, take care of what you really need first. You might even have enough for that holiday or car you've always dreamed about. This section is about setting up a budget and managing your money.

Budget: Plan for money you need or have.

Budgeting tips:

★ Find out how much money you actually take home each month (approximately four weeks). Net monthly income (total pay less tax etc.) = 4 × weekly take-home pay.

★ Deduct from your net monthly income all fixed expenses (those which are the same or nearly the same each month such as rent, rates, gas, electricity, fares).

★ Adjust your flexible expenses (those which may vary monthly or are not really needed each month) based on the money you have left after deducting the fixed expenses.

★ Decide what you need and how much to spend on such things as food, clothing and entertainment.

★ Don't forget to include some money for savings and emergencies.

What would you do if you were the people in Ex.1 and Ex. 2? Fill out the budget sheet for each person. First find the total amount of fixed expenses. Then change the flexible expenses so that each person can save money.

1. Linda earns £106.00 a week as a secretary. Her net income per week is £78.50. Here is a list of her expenses last month.

Lunches	£28.00
Cinema	8.00
Rent	87.00
Hairdresser	9.40
Telephone	7.30
Electricity	12.80
Clothes	48.60
Fares	24.00
Food	56.00
Loan Payment	12.00
Miscellaneous	20.90

Linda wants to save. Help her to decide which expenses to cut down. Net monthly income is £78.50 × 4 = £314

Fixed expenses

Rent	£ 87.00
Telephone	7.30
Electricity	12.80
Fares	24.00
Loan payment	12.00
Total fixed expenses A.	£ 143.10
Balance B.	£ 170.90

Flexible expenses

....	£ ....
....	....
....	....
....	....
....	....
....	....
Total flexible expenses C.	£ ....
Savings (subtract C from B.)	£ ....

2. Tim's job at the record shop pays £94.00 a week. His take-home pay is £68.50. Here is a list of Tim's expenses last month.

Entertainment £38.00

Rent 52.00

Telephone 5.20

Electricity 11.30

Car payment 20.00

Food 48.00

Petrol & repairs 41.60

Clothes 39.00

Sports Club subscription 4.50

Miscellaneous 14.40

Tim wants to go to night school. He needs to save £34 a month. Help him work out a budget. Net monthly income is .... × 4 = ...

Fixed expenses

.... £ ....

....

....

....

....

Total fixed expenses A. £ ....

Balance B. £ ....

Flexible expenses

.... £ ....

....

....

....

....

....

Total flexible expenses C. £ ....

Savings £ ....

Frank and Ernest   **by Bob Thaves**

On Your Own

Now it's your money you must budget. How much do you receive each month? Remember, take care of what you really need first. Then use what's left for your other expenses. If you wish to save for something you really want, you can do it! Work out your budget and stick to it!

Net monthly income .... × 4 = £ ....

Fixed expenses

.... £ ....

....

....

....

....

Total fixed expenses £ ....

Balance £ ....

Flexible expenses

.... £ ....

....

....

....

....

Total flexible expenses £ ....

Savings .... £ ....

Renting a Place to Live

Is looking for a place to live like decoding a secret message? All the codes in the advertisements are about what is offered and how much it will cost you. If you take some time to learn exactly what the advertisements say, finding the actual cost of renting a home is not a great mystery!

Which places should the persons in Ex. 1 and Ex. 2 rent? Read the facts about them and help them to choose. Remember to include public transport costs in making your decisions.

Fact box ▶

The following abbreviations and words are often used in ads:

b/sit – bedsitter
bth – bath
bed – bedroom
kit – kitchen
rec – reception/living room
dbl – double
deposit – money to cover any damage you may
      cause, returnable at the end of your lease.

s/c – self-contained
c.h. – central heating
f.f. – fully furnished
p.w. – per week
lease – rent agreement
p.c.m. – per calendar month

1. Jenny Wilson's weekly take home pay is £74. She wants to rent either flat A or flat B. She can walk to work from flat A, but electricity will cost her at least £2.70 a week. She has to catch a bus from flat B at 35p a journey for 10 trips a week. The most Jenny can afford is £28 per week.

   A. Close city. S/c flat. 1 bed, kit, bath, rec. £24 p.w. Tel. 406-6282
   B. Forest Hill. 1 bed flat. ftd. kit, lounge, c.h. £24 p.w. inclusive. Tel. 316-0243

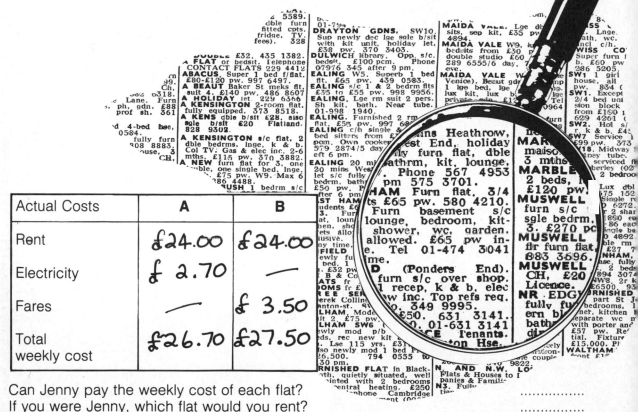

Actual Costs	A	B
Rent	£24.00	£24.00
Electricity	£2.70	—
Fares	—	£3.50
Total weekly cost	£26.70	£27.50

Can Jenny pay the weekly cost of each flat?
If you were Jenny, which flat would you rent?

2. Jimmy Read's weekly take-home pay is £94. His budget for rent and fares is 35% of his weekly income. How much is this? .... Electricity would cost him about £3 for flat B. The cost of a return train ticket to work from flat A is 85p and Jimmy makes at least 8 return trips a week. He could walk to work from flat B.

A. Southside – 1 bed flat, c.h. £26 p.w. including elec. Tel. 332-0463
B. Hunting Park – 1st floor mod. flat, £29 p.w. plus elec. Tel. 482-3214

Actual Costs	A	B
Rent	....	
Electricity	....	
Fares	....	
Total weekly cost	....	

Can Jimmy afford the total weekly cost for each flat? ....
If he must rent one of them, which should he choose? ....
Why? ....

On Your Own

You have a choice between two flats.
Flat A is on a bus route and a single journey to work will cost 20p. Flat B requires two journeys of 25p each to get to work. The average cost of electricity is £2.80 a week.

A. 1 bed. flat Wood Green f.f. near park, available immediately. £27 p.w. plus electricity. Tel. 437-2484
B. 1 bed.flat Westfield, lge. room, mod kitchen, £25 p.w. plus electricity. Tel. 470-2737

Actual costs	A	B
Rent	....	....
Electricity	....	....
Fares	....	....
Total weekly cost	....	....
Which would you rent?		....

Home Ownership

You want to buy a home and need to borrow some money. Where's the best place to go for a mortgage? How much can you borrow? How much will it cost? This section will help you to understand the types of mortgages that are available.

Mortgage: The loan made to you against the security of your home.
There are two main methods of repaying the loan.

a. Repayment mortgage: Payment of a monthly instalment of capital and interest for an agreed period of years.

b. Endowment mortgage: Repayment of the loan is made from the proceeds of a life assurance policy on maturity. You pay the interest on the loan monthly to the building society plus an assurance premium to an assurance company.

Rate: The rate of interest payable on the loan (this may vary during the period of the loan).

Term: The period of time which you have to repay the loan.

MIRAS: Mortgage Interest Relief at Source. This means the building society claims your tax relief from the Inland Revenue and you have a lower net monthly payment to the Building Society.

Most home buyers have their mortgage from a building society. The amount of money the building society will lend you depends upon your income. If there are two people both incomes are taken into account.

Table 1. shows the monthly cost (in £) for each £1000 of mortgage.

1.

Length of loan in years	Rate of interest		
	10%	12%	14%
15	8.46	9.24	10.07
20	7.29	8.16	9.09
25	6.69	7.63	8.63

Table 2. shows the lending policies of some building societies.

2.

Society	Multiple of income lent	
	One income	Two incomes H – higher L – lower
Abbey National	2½	2½H + 1L
Bradford & Bingley	2	2H + 1L
Gateway	2¼	2¼H + ½L
Halifax	2½	2¼H + 1L
Leeds	2½	1¾H + 1¾L
Nationwide	2½	2½H + 1L
Woolwich	2¾	2¾H + 1L

Use the FACT BOX and TABLES to answer questions 1-8. Exs·1 and 2 are done for you.

1. John Stewart has a mortgage of £7000 for twenty years at 12%. What are his monthly repayments?

   Monthly repayments = 7 × 8.16 (from Table 1)

   = £57.12

2. Peter and Susan Lewis want to buy a house. Peter earns £6000 a year and Susan £4000. What is the maximum mortgage they can obtain from the Gateway?

   Maximum mortgage = 2¼ × higher income + ½ × lower income

   = 2¼ × 6000 + ½ × 4000

   = 13500 + 2000

   = £15500

3. Mary Blake has a mortgage of £8000 for twenty-five years at 14%. What are her monthly repayments? ....

4. What is the maximum mortgage the Abbey National will give Brian and Joan Williams, if Brian earns £5000 a year and Joan earns £7000 a year? ....

5. What is the minimum annual income Steven Dawson must have if he wishes to obtain a mortgage of £30,000 from the Nationwide? ....

6. Sharon Baxter's income is £6800 a year. Will she be able to obtain a £16500 mortgage from the Halifax?

7. Ian Kelly and his brother Sean want to buy a flat together. Ian earns £6500 a year and Sean £4200. How much will the Woolwich lend them? ....

8. Tony Scott has a £12000 mortgage for twenty years at 14 % The mortgage rate is lowered by 2 % from 14 % to 12 %. How much less does Tony have to pay each month?

....

On Your Own

When buying a house there are many additional costs you have to budget for.
Find out all you can about: stamp duty, surveyors' fees and solicitors' charges.

Are You Covered?

The cost of motoring is increasing every year. Insurance accounts for a large part of the cost of running a car. By law all motor vehicles must be covered by insurance. This section will help you to understand how the cost of insuring your car is worked out.

Fact box

Basic Premium The cost of insuring a vehicle, before any deductions. This depends on such factors as make and age of vehicle, size of engine, district where you live, driver's age and record, and the use of the vehicle (business, pleasure etc.).

Payable Premium The actual amount paid for insuring your vehicle. Payable premium = Basic premium less no claims bonuses and other specified deductions.

No Claims Bonus If your car has been insured for one year and you have not made any claims against the insurance company, then your premium for the next year is reduced.

No Claims Bonus Scale

Insurance Period	Reduction on Basic Premium
After one year no claims	30%
After two consecutive years no claims	40%
After three consecutive years no claims	50%
After four or more consecutive years no claims	60%

Other deductions include 10% if only two particular people are allowed to drive the vehicle and 25% if you are in a certain employment category (civil servants, etc.).

Insurance Group Each type of car is rated in an insurance group from 1 to 9. The higher the group, the more expensive the basic insurance premium.

Shown below is an example of a vehicle insurance schedule.

1.

Make	c.c.	Year of Manufacture	Type of Body	Registration	Value	Group	Area
Datsun Cherry Coupé	1171	1980	Saloon	JLR 752V	£3850	4	1

```
Vehicle          JLR 752V
Cover            Comprehensive
Basic Premium        £390.00
Driver        − 10%    39.00  (2 named drivers only)
Total                351.00
S.R.          − 25%    87.75  (job category)
Total                263.25
N.C.B.        − 60%   157.95  (full no claims bonus)
Total                105.30
Premium payable      £105.30
```

Using the Fact Box and the steps shown in Exs. 1 and 2 answer questions 3-6.
Find the premium payable on the following cars.

2. Car A Basic Premium £480.00
    Two named drivers— 10% 48.00
                  — (0.1×480)
    Total 432.00
    Five years N.C.B. — 60% 259.20
                  — (0.6×432)
    Total £172.80
    Premium payable £172.80

3. Car B Basic Premium £210.00
    S.R. (job categ.) ....
    Total ....
    Three years N.C.B. ....
    Total ....
    Premium payable ....

4. Car C Basic Premium £306.00
    Two named drivers ....
    Total ....
    Five years N.C.B. ....
    Total ....
    Premium payable ....

5. Car D Basic premium £425.00
    Two named drivers ....
    Total ....
    S.R. (job categ.) ....
    Total ....
    Two years N.C.B. ....
    Total ....
    Premium payable ....

If you have an insurance claim during the year, the following year you drop two places on the N.C.B. scale.
60% N.C.B. is reduced to 40%
50% N.C.B. is reduced to 30%
40% and 30% is lost for that year.
If more than one claim is made in a year you lose all your N.C.B. for the next year.

---

6. If each of the owners of cars B, C and D have one claim each during their period of insurance cover, find the increase in payable premium for the next year.

On Your Own

Find out what other specified deductions insurance companies will make. What is a "Green Card", and where is it used? Choose a car you would like to own and obtain an insurance quotation for it. What is the difference between Comprehensive and Third Party insurance?

All About Credit

You have just discovered that you do not have enough money to buy something you really need. Should you borrow money from a bank or a finance company? Should you use your credit card to spread the payments over a number of months? In all these cases, you are using credit. This section is about credit and what it costs.

When buying on credit, you usually have to pay extra for the goods. If you buy the items in Exs.1-4 on credit, how much more will you pay?

1. **Colour television**
   £350 cash or £50 deposit and £28 per month for twelve months.

Deposit	£50
Total monthly payments	£336 (28 × 12)
Total amount of payments	£386
Less cash price	£350
Cost of credit	£36

2. **Record player**
   £5 monthly for 1 year or £52.95 cash.

Total amount of payments	£.................
Less cash price	£.................
Cost of credit	£.................

3. **Freezer**
   £220 cash or twelve monthly payments of £21

Total amount of payments	£.................
Less cash price	£.................
Cost of credit	£.................

4. **Stereo system**
   £285 cash or £80 deposit and twenty-four monthly payments of £10

Deposit	£.................
Total monthly payments	£.................
Total amount of payments	£.................
Less cash price	£.................
Cost of credit	£.................

To find which loan is the cheapest, you take the interest percentage rate per month. Use the method shown in the example to find the rate of interest paid for one month on each loan in Ex. 5.

Example: What is the monthly interest rate on a £300 loan for five months with a total interest charge of £30?

Monthly rate =

$$\frac{\text{Interest charge}}{\text{Amount of loan}} \times \text{number of months}$$

$$= \frac{30}{300 \times 5}$$

$$= 0.02$$

Percentage rate =
$$0.02 \times 100 = 2\% \text{ per month}$$

5. Which of these loans has the lowest rate of interest?

   a. Credit card loan of £500 with total interest charge of £50 fully paid after ten months.

   b. A finance company loan of £400 with total interest charge of £40 fully paid after five months.

   c. A savings bank personal loan of £600 with total interest charge of £54 fully paid after twelve months.

Do you know what is sometimes called "plastic money"? If you said "a credit card", you were right!

You have to be a careful consumer to use a credit card wisely. Use the following information and the Barclaycard statement to answer Exs. 6-8

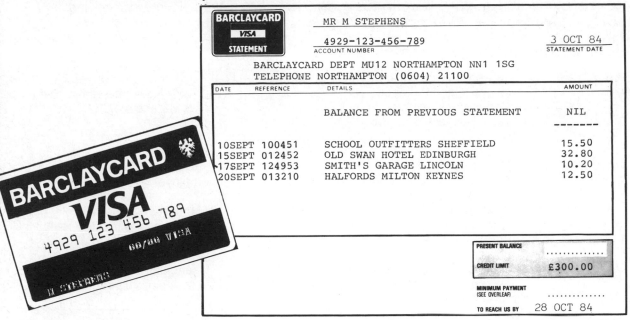

BARCLAYCARD VISA STATEMENT	MR M STEPHENS		
	4929-123-456-789 ACCOUNT NUMBER		3 OCT 84 STATEMENT DATE
	BARCLAYCARD DEPT MU12 NORTHAMPTON NN1 1SG TELEPHONE NORTHAMPTON (0604) 21100		

DATE	REFERENCE	DETAILS	AMOUNT
		BALANCE FROM PREVIOUS STATEMENT	NIL

10SEPT	100451	SCHOOL OUTFITTERS SHEFFIELD	15.50
15SEPT	012452	OLD SWAN HOTEL EDINBURGH	32.80
17SEPT	124953	SMITH'S GARAGE LINCOLN	10.20
20SEPT	013210	HALFORDS MILTON KEYNES	12.50

PRESENT BALANCE ....

CREDIT LIMIT £300.00

MINIMUM PAYMENT (SEE OVERLEAF) ....

TO REACH US BY 28 OCT 84

Credit Charges
★ If payment is received by the due date no credit charge is payable.

★ If you do not wish to pay the full amount, you must make a minimum repayment, equal to 5% of the balance or £5 (whichever is the greater).

★ If you do not pay the full amount, interest is charged on the outstanding balance at the rate of 2.25% per month on a daily basis.

6. What is the present balance on the Barclaycard statement? ....

7. What would be the minimum payment on this statement? ....

8. If you make this minimum payment, how much is outstanding on this statement?

....

On Your Own

If you do not have a credit card yet, you will probably apply for one soon. Obtain a Barclaycard information sheet and application form from a bank. Try filling it in.

Find out the difference between credit cards (e.g. Barclaycard), charge cards (e.g. Diners Club, American Express) and Budget Account cards (e.g. Wallis Shops, Boots, etc.).

Your Income Tax

Once you begin earning money, you must complete an Income Tax Return form. This is needed to allow the tax office to work out the correct amount of income tax you will have to pay. Most people complete a tax form every year, to allow for changes in earnings and personal circumstances. This section will help you to understand how income tax is worked out.

Fact box

P.A.Y.E. tax:	Pay-As-You-Earn tax. Money for the Inland Revenue Department deducted from each pay packet by your employer.
Taxable income:	Total income minus tax allowances. The amount of your income on which you must pay tax.
Tax allowances:	Personal and other admissible charges that a person is entitled to claim on his or her tax return (expenses, etc.).
Tax code:	Your tax code is based on your total tax free allowances. (Allowances of £3622 = tax code 362.)
P60 form:	A statement of your earnings and tax deducted by your employer.
Tax rate:	The percentage of taxable income paid in tax. The rate of income tax for each current year is announced by the Chancellor of the Exchequer in his budget speech.

The rates of tax and bands of taxable income for 1984–1985 were:

		£
Basic rate	30%	1 – 15400
Higher rate	40%	15401 – 18200
	45%	18201 – 23100
	50%	23101 – 30600
	55%	30601 – 38100
	60%	over £38100

Martin Young is a surveyor working for Barnes Construction Company, High Road, Leeds. His wife, Angela, works as a secretary for B & W Carpets Ltd, East Lane, Leeds. Martin earns £8540 a year and Angela £5723. Martin has an expense allowance of £300 a year for travel in connection with his work. Angela pays £15 in union subscriptions, and Martin pays £45 a year to the surveyors' professional organisation. Martin claims £80 for protective clothing.

Using the details of the Youngs' income and allowances, complete parts A and B of the Tax Return form.

A

INCOME: Year ended 5 April 1981

	See Note	Details	Amount for year	
			Self £	Wife £
EARNINGS		Own full-time employment. *You need not enter details or pay relating to your own full-time employment but you must enter tips below.*		
	1	Wife's employment *Employer's name and address Works No. (if any)*		
	2	All other earnings *Type of work*		
		Name and address of anyone for whom work done		
	3	Tips and incidental receipts from ALL sources		
		If you or your wife received a taxed sum from trustees of an **approved profit sharing scheme** tick "√" here	▶	*(see also note 31)*
		If the duties of your employment were performed wholly or partly outside the United Kingdom tick "√" here	▶	

B

		Expenses against earnings			Self £	Wife £
	4	If a fixed deduction applies tick "√" here ▶	Self	Wife		
		Where no fixed deduction applies state nature and amounts of expenses				
	5	Fees or subscriptions to professional bodies Name of professional body				

Follow the steps in Exs. 1 and 2 to calculate the amount of income tax paid in one year.

1. David Newton has a salary of £10740. He has allowances of: Personal £2005, Expenses £85, Union Subscriptions £25. Find the amount of tax he pays in one year.

Allowances: Personal................................. £2005.00
            Expenses............................. £85.00
            Union..................................... £25.00 +
                  Total....................... £2115.00

Taxable income $=$ total income $-$ allowances
$= £10740 - £2115$
$= £8625$

Tax payable at basic rate of 30% $= 0.3 \times 8625$
$= £2587.50$

2. Geoff Adams earns £20860 a year. His allowances are: Personal £2005, Expenses £460, Housekeeper £150, Alimony £2000. Calculate the amount of tax he pays.

Allowances: Personal................................. £2005.00
            Expenses............................. £460.00
            Housekeeper........................ £150.00
            Alimony................................. £2000.00 +
                  Total....................... £4615.00

Taxable income at basic rate of 30% $= £15400$
Tax payable at basic rate $= 0.3 \times 15400$
$= £4620$
Taxable income at higher rate of 40% $= £845 \, (16245 - 15400)$
Tax payable at higher rate of 40% $= 0.4 \times 845$
$= £338$
Total tax payable $= £4620 + £338$
$= £4958$

Taxable income $=$ total income $-$ allowances
$= £20860 - £4615$
$= £16245$

3. Find the amount of income tax paid by Carol Ford if she has an annual salary of £8295 and allowances of: Personal £2005, Expenses £42, Dependent relative £145.

4. Paul Roberts earns £18250 a year. His allowances are: Personal £2005, Expenses £54. Find the amount of income tax Paul pays in one year.

5. Paul Roberts from Ex. 4 gets married. His salary remains the same at £18250 per annum but his allowances change to: Personal £3155, Expenses £54. How much tax does he have to pay now?

6. What is the difference between Paul Roberts' tax in Exs. 4 and 5?

On Your Own

Get a Tax Return form from your local Tax Office and fill in as much information as you can. Tax codes have a letter after a three digit number, e.g. 246L. Find out what these letters mean.

Looking Back

1. Fill out this deposit slip with four £5 notes, ten £1 notes, £2.00 in silver and £1.45 in bronze coin.

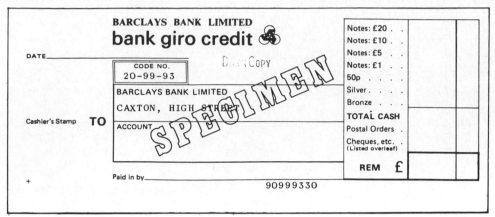

2. Write a cheque for £43.26 to the Electricity Board for payment of an electricity bill.

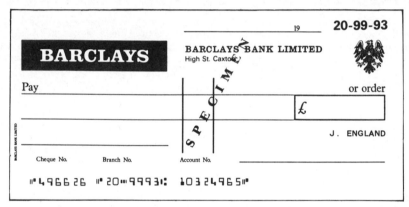

3. Show the deposit and cheque from Exs. 1 and 2 on the cheque counterfoil.

4. You open a Building Society account with a deposit of £300. If you keep the money in the account for 6 months how much interest will it earn at a rate of 8% per year? ......

5. Your net weekly income is £95.00. You allow £40 a week for rent and fares. Electricity usually costs £5 a week. Choose between these two flats. You can walk to work from flat A. The journey to work from B costs 22p and you make at least 10 trips a week. Which flat should you rent?
   Flat A.   North Road. Walk to work. Lge bdrm with kit. & bth. £35 plus elec.
   Flat B.   Hurst Park. S/c flat 1 bdrm. kit. & bth. £35 p.w. inc. elec.

Skills Survey

You have seen how useful your maths skills are in managing your money. The exercises in this section will help to sharpen your skills.

1. Write these amounts in words and figures as you would for a cheque.

£6.00 ....

£58.34 ....

£101.50 ....

£1200.00 ....

2. Fill in the balance line after each entry.

ACCORD
BUILDING SOCIETY

Name .....Winston Clive.............................

.............HUGHES.............................

Roll no.   2/   10824 354 – 9

Branch   High St. Newtown

Date	Cashier's initials	Details	Withdrawals	Receipts	Balance
		Brought Forward			1245.68
3/8/84	J.O.	Fifty pounds ch.		50.00	
2/9/84	S.B.	One hundred & sixty two pounds		162.00	
1/10/84	S.B.	Eighty pounds & 74 pence		80.74	
9/11/84	C.A.	One hundred & twenty pounds	120.00		
4/12/84	T.L.	Fifty pounds & 45 pence	50.45		

3. What is the total yearly cost of these monthly payments?

Amount: Amount: Amount: Amount:
£225.00   £68.13   £170.35   £96.42

Yearly cost: Yearly cost: Yearly cost: Yearly cost:

....

4. What is the monthly rate of interest for this loan?

Amount borrowed: £500 for five months
Interest: £50
Rate of interest: ....

5. How much interest will you earn per year for each deposit?

£550 at 12% per year: Amount of interest ....

£2250 at 11½% per year: Amount of interest ....

6. You earn £140 a week and your take-home pay is £107. How much is deducted each month? ....

How would you fill out this budget sheet if your usual monthly expenses are:

Mortgage £140
Clothing £50
Food £60
Car payment £45
Gifts £21
Electricity £25
Telephone £18
Cinema and lunches £60

Net monthly income: £.................

Fixed expenses:	
.... £.............
....
....
....
Total fixed expenses £................
Balance £................

Flexible expenses:	
.... £.............
....
....
....
....
Total flexible expenses £................
Savings £................

7. You can pay for a radio cassette player with £39.95 in cash. Instead, you decide to pay £4.60 a month for ten months. How much more do you have to pay? ....

8. Your total income is £6350. Your allowances are: Personal £2005. Expenses £43, Union Fee £20. What is your taxable income?

....

Branching Off

Banks offer several types of savings plans. Get a brochure from your local bank and decide which plan is best for you.

There are many kinds of insurance — life, house, fire and theft, car, etc. Find out all you can about insurance and decide which policies might be suitable for you. However, don't let an insurance agent talk you into buying a policy you don't need!

Unit 4

Do you often ask yourself, "How can I make more money?" If you do, this unit is for you!

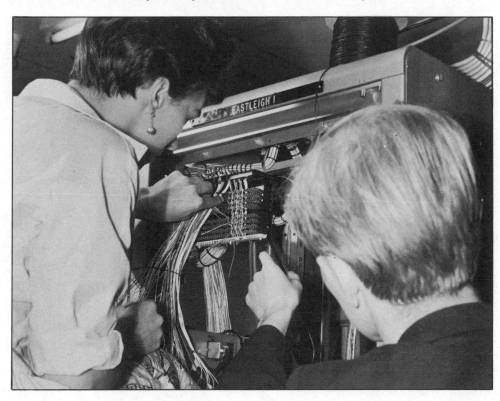

Maths at Work

The Best Paid Job

Do HELP WANTED advertisements tell you exactly how much you will be paid when you get a job? This section will help you work out the take-home pay you can expect from casual jobs described in advertisements.

Fact box

Hr day = hours worked in a day.
Days wk = days worked in a week.
Rate = pounds per hour worked.
Gross pay = hourly rate x total hours worked.
P.A.Y.E. = Pay-As-You-Earn income tax. Money deducted from pay by the Inland Revenue. Higher gross pay usually means a higher rate of tax.

Net pay (take-home pay) = gross pay minus deductions.
Casual job = temporary work on an hourly rate of pay.

GARDENER £2.50/ hr

9.30–3.30, five days a week. Light work, inc. lawns. Must be reliable.
Phone 704-8742.

FORECOURT ATTENDANT £2.10/hr

Five days, 9 a.m.-1 p.m.
General duties.
Apply: Al's Garage,
York Street, City.

AL'S GARAGE

BUILDER'S LABOURER
£3.60/hr
7.30 a.m.-1.30 p.m., six days a week.
On-site work. Must be experienced.

Topline Homes,
Beeton Road,
London NW 5.

Prompt Taxis

MINICAB DRIVER

£3.30/hr
Fri-Sun., 1 p.m.-7 p.m.
Must have clean licence.
Tel. 725-4532

Fill in the missing amounts on the weekly pay slip for each job. Follow the steps used in the example.

1. GARDENER

Hr/day 6	P.A.Y.E. tax: £15.75
Days/wk 5	Social security £5.25
Total hours 30	Other deductions —
	Total deductions
Rate £2.50	(B) £21.00
Gross pay (A)£75.00	Net pay (C)£54.00

(A) £2.50   (B) £15.75   (C) £75.00·
　　　30 ×　　　　　5.25 +　　　　21.00 −
　　£75.00　　　£21.00　　　　£54.00

2. FORECOURT ATTENDANT

Hr/day	P.A.Y.E. tax: £6.30
Days/wk	Social security £2.84
Total hours	Other deductions —
	Total deductions
Rate	
Gross pay	Net pay

3. BUILDER'S LABOURER

Hr/day	P.A.Y.E. tax: £31.45
Days/wk	Social security £9.52
Total hours	Other deductions (union) £0.80
	Total deductions
Rate	
Gross pay	Net pay

4. MINI-CAB DRIVER

Hr/day	P.A.Y.E. tax: £10.24
Days/wk	Social security £4.41
Total hours	Other deductions (insurance) £2.10
	Total deductions
Rate	
Gross pay	Net pay

On your Own

Find HELP WANTED advertisements in your local paper. Is there a casual job that might suit you? Calculate take-home pay if deductions are 25% of gross pay.

Working Time

What time do you come to work? When do you leave? Your answers could mean money! Your salary or wage often depends upon the amount of time you spend working.

This section is about measuring your time at work.

1 day = 24 hours
1 hour (hr) = 60 minutes (min)
★ Any amount of time more than 59 minutes should be changed into hours and minutes by dividing the minutes by 60.

```
         1 hr 25 min
60 ) 85 min
       60
       25
```

Here are examples of how time is calculated:

★ Adding time:

```
  1 hr 25 min
+ 2 hr 55 min
  3 hr 80 min
```

```
            1 hr 20 min          3 hr
60 ) 80                  + 1 hr 20 min
       60                  4 hr 20 min
       80
```

★ Subtracting time:

```
                        - - - - - -(60 + 15) min
  7 hr 15 min  = 6 hr 75 min
- 5 hr 45 min  - 5 hr 45 min
                 1 hr 30 min
```

★ Multiplying time:

```
  5 hr 45 min
       × 5
25 hr 225 min
```

```
             3 hr 45 min          25 hr
60 ) 225                  + 3 hr 45 min
      180                   28 hr 45 min
       45
```

★ Dividing time:

```
     5 hr       9 min
7 ) 36 hr     3 min
     35
      1 hr = 60 min
             63 min
             63
             00
```

TRY IT!

Add:
```
  5 hr 45 min
+ 4 hr 20 min
```

Subtract:
```
  8 hr 25 min
- 4 hr 45 min
```

Multiply:
```
1 hr 25 min
       ×6
```

Divide:
```
5 ) 21 hr 15 min
```

1. The following chart shows the amount of time the people employed by Best Bookshop worked per day. Find the total time each employee spent at work:

	Crawford	Roberts	Schmidt
MONDAY	7 hr 30 min	8 hr 40 min	6 hr 45 min
WEDNESDAY	5 hr 45 min	6 hr 15 min	7 hr 35 min
FRIDAY	7 hr 10 min	5 hr 50 min	8 hr 20 min
TOTAL TIME			

2. Don't forget lunch breaks! You don't always get paid for them. What is the actual working time for each of these employees at the Tip-Toe Shoe Shop?

	Papanis	Stevens	Hannigan
TOTAL TIME AT WORK	37 hr 30 min	40 hr 25 min	38 hr 45 min
LUNCH	2 hr 30 min	5 hr	3 hr 50 min
TOTAL TIME			

3. These employees know how much time they worked on their first day. They want to know how much time they might be able to put in each week. Calculate the weekly time for each employee.

	McKay	Poulos	Cooper
TIME IN ONE DAY	8 hr 10 min	7 hr 30 min	6 hr 45 min
NUMBER OF DAYS AT WORK	4	5	6
TOTAL TIME FOR ONE WEEK			

4. Sometimes you may want to find out what your average time is for each working day. Find the average for each of these employees at Sunshine Appliances. Divide the total time by the number of days.

	Lester	Jardine	Clyde
TOTAL TIME FOR ONE WEEK	36 hr 15 min	37 hr 30 min	25 hr 20 min
NUMBER OF DAYS AT WORK	5	6	4
AVERAGE TIME PER DAY			

On Your Own

Now you're a time expert. Use your skills to calculate the average time you spend on each of your daily activities.

Overtime

How would it feel if your pay packet was larger than you expected? Great! Yes, it can happen - if you are asked to work overtime. This section will help you understand overtime pay and how it adds to a normal salary.

Calculate the gross earnings of the following employees based on their time cards. Follow the worked example.

BEST BOOKSHOP

WEEK ENDING 13 October
NAME Carol Scott

DAYS	IN	OUT	IN	OUT	Daily Total	Ord. hours	Overtime (T½)	(DT)
M.	8–15	12–00	1–00	5–15	8	8		
Tu.	8–00	12–00	1–00	5–00	8	8		
W.	8–10	12–00	12–40	5–50	9	8	1	
Th.	8–00	12–00	12–30	6–30	10	8	2	
Fri.	8–15	12–15	1–00	4–00	8	8		
Sat.								
WEEKLY TOTAL					43	40	3	

TOTAL ORD. HOURS   40

(T½) **3**.. hr x 1.5   4.5

(DT) ..... hr x 2   _____

TOTAL EQUIVALENT HOURS   44.5

RATE   £1.80

GROSS EARNINGS   £80. 10

60

fit well
shoe shop

WEEK ENDING **4 MAY**

NAME **Mike Peters**

DAYS	IN	OUT	IN	OUT	Daily Total	Ord. hours	Overtime (T½)	(DT)
M.	8–30	1–00	2–00	5–30	8			
Tu.	8–20	1–20	2–00	6–00	9			
W.								
Th.	8–30	1–30	2–00	7–00	10			
Fri.	8–30	12–30	1–30	5–30	8			
Sat.	8–00	12–00	1–00	6–00	9			
WEEKLY TOTAL								

TOTAL ORD. HOURS _____

(T½) ..... hr x 1.5 _____

(DT) ..... hr x 2 _____

TOTAL EQUIVALENT HOURS _____

RATE **£1.80**

GROSS EARNINGS _____

Pawson & Son engineers

WEEK ENDING **24 MARCH**

NAME **Dennis O'Neil**

DAYS	IN	OUT	IN	OUT	Daily Total	Ord. hours	Overtime (T½)	(DT)
M.	8–00	12–00	1–00	5–00				
Tu.	8–00	12–00	1–00	6–00				
W.	7–30	12–30	1–30	6–30				
Th.	7–30	12–30	1–30	6–30				
Fri.	7–00	12–00	1–00	7–00				
Sat.								
WEEKLY TOTAL								

TOTAL ORD. HOURS _____

(T½) ..... hr x 1.5 _____

(DT) ..... hr x 2 _____

TOTAL EQUIVALENT HOURS _____

RATE **£2.00**

GROSS EARNINGS _____

On Your Own

Imagine you are being paid to go to school. Fill out the time card for a week at school. Assume your hourly rate is £1.10. Calculate the total hours worked per week and gross earnings.

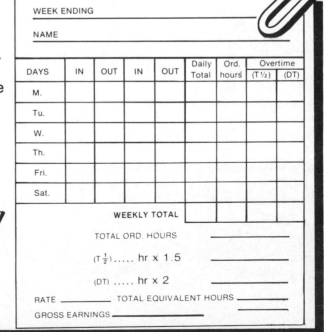

WEEK ENDING _____

NAME _____

DAYS	IN	OUT	IN	OUT	Daily Total	Ord. hours	Overtime (T½)	(DT)
M.								
Tu.								
W.								
Th.								
Fri.								
Sat.								
WEEKLY TOTAL								

TOTAL ORD. HOURS _____

(T½) ..... hr x 1.5 _____

(DT) ..... hr x 2 _____

RATE _____ TOTAL EQUIVALENT HOURS _____

GROSS EARNINGS _____

Earning by the Piece or by Commission

What rewards do you get for working hard? If your pay is based on the number of things you make or sell, the rewards for hard work are easy to see. You earn more when you make or sell more! This lesson will help you to understand how piecework earnings and commission on sales are calculated.

Fact box →

Piece rate: Amount of money earned for each piece made.

Piecework earnings: Piece rate × number of pieces made.
Example: A jeweller earns £1.20 for each piece of jewellery. How much is earned for making fifty-six pieces?
Piecework earnings = £1.20 × 56
= £67.20

Commission: A percentage of a salesperson's total sales.
Example: Suppose you sell cars at 5% commission. How much will you earn if you sell a car for £4000?
Commission = 5% of £4000
Commission = $\frac{5}{100} \times \frac{4000}{1}$
= £200

Read the facts about each person carefully. Then calculate his or her earnings.

1. Susan makes canvas bags at a piece rate of 75p. If she makes ninety-five bags in one week, what is her weekly pay?

   Susan's
   earnings   .... × .... = ...

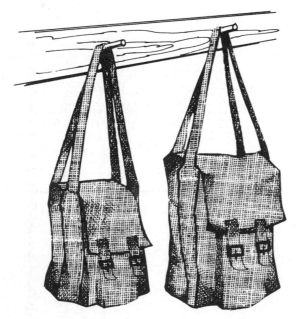

2. Mark makes belts in three sizes. The piece rates for each size are: small = 50p; medium = 75p; large = £1.00. Calculate Mark's total earnings on the chart.

Size	Number of belts made	Piece rate	Earnings
Small	25		
Medium	29		
Large	27		
TOTAL		/////	

3. Patricia Collins sold a house for £24,000. If her commission was 6%, how much did she earn?

Commission = 6% of £ ....
= .... × ....
= ....

4. Tony Evans, a travel agent, earns different commissions for different types of travel. Calculate his earnings on the sales listed on this chart.

Type of travel	% Commission	Amount of sale	Earnings
Regular trips	7%	£455	
Charters	5%	£699	
Escorted Tours	10%	£550	
Transportation & Hotel Packages	11%	£851	

5. Suppose you earn 20p for each record that you sell for £4.00. What percentage commission are you being paid?

$$\text{Percentage commission} = \frac{\text{earnings}}{\text{sales}} \times 100$$

$$= \frac{£0.20}{£4.00} \times 100$$

$$= \frac{20}{400} \times 100$$

$$= .... \times ....$$

$$= .... \%$$

6. Calculate the percentage commissions on the following:

Earnings	Amount of sale	Percentage commission
£0.02	£0.50	....
£450	£9000	....
£1.80	£15	....
£24	£300	....

On Your Own

When you sell a house through an estate agent you must pay a commission. Find out how this commission is calculated in your area. List the selling prices of three houses from the HOUSES FOR SALE column in your local newspaper and calculate the commission payable.

	Sale price	Commission
House A	....	....
House B	....	....
House C	....	....

What is Profit? What is Loss?

When you buy a bicycle for £16.00 and sell it for £20.00, your profit is £4.00. But suppose you spend £5.00 for advertisements before you sell the bicycle. Then you have made a loss of £1.00. This section will help you to understand profit and loss in business.

Total sales: The sum of the amounts that the seller receives from customers.

Cost of goods sold (cost price): The amount that the seller pays for the things sold.

Selling price: The amount the customer pays for the goods.

Operating expenses: The sum of amounts paid for doing business (rent, electricity, telephone, office supplies, salaries, advertising and others).

Gross profit: Total sales minus cost of goods sold.

Net profit: Gross profit minus operating expenses.

Net loss: The difference between gross profit and operating expenses, if the expense amount is greater than the profit.

Stock value: Number of goods for sale × unit cost.

Read the following facts and answer the questions.

1. The roller skates that you bought for £8 were sold for £9.50. What was your gross profit?   ....

2. You bought a plain T-shirt for £1.99. The iron-on letters that you put on the shirt cost you £1.10. How much should you sell the T-shirt for to earn a profit of £2.00?

   Cost of plain T-shirt   ....

   Additional cost of letters + ...

   Cost of T-shirt for sale   ....

   Profit   + ...

   Selling price   ....

3. When you tried to sell the T-shirt at your selling price, nobody wanted to buy it! So you sold it for £2.50. Did you make a profit or a loss?

   Cost of T-shirt for sale   ....

   Amount paid to you   ....

   Difference   ....

   Is this a profit or a loss? ....

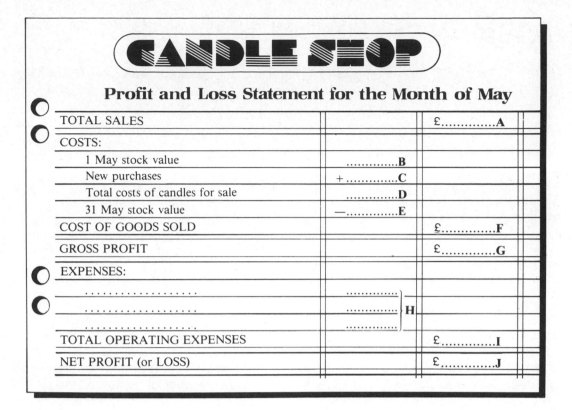

CANDLE SHOP

Profit and Loss Statement for the Month of May

TOTAL SALES		£.............A
COSTS:		
1 May stock value	.............B	
New purchases	+ C	
Total costs of candles for sale	.............D	
31 May stock value	— ....E	
COST OF GOODS SOLD		£.............F
GROSS PROFIT		£.............G
EXPENSES:		
....	....	
....	.... } H	
....	....	
TOTAL OPERATING EXPENSES		£.............I
NET PROFIT (or LOSS)		£.............J

Prepare a **PROFIT & LOSS STATEMENT** for this business. Read the facts for each letter carefully.

A. The weekly sales in May were:

First week:	£155.50
Second week:	£186.75
Third week:	£195.00
Fourth week:	£175.25
Total sales	£

B. On May 1, there were 1500 candles in the shop at £0.05 each.

1500 × £0.05 =   £.................

C. The new candles bought in May cost £200.

D. Add **B** and **C**.

E. On May 31, there were 2000 candles in the shop at £0.05 each.

2000 × £0.05 =   £.................

F. Subtract **E** from **D**.

G. Subtract **F** from **A**.

H. To run the shop, the owner paid £200 rent, £100 for ads and £95.50 for office supplies.

I. Add the amounts in **H**.

J. Subtract **I** from **G**.

Note: If expenses are greater than the gross profit, the difference is **LOSS.**

On Your Own

Suppose you want to earn money by making model cars. Choose a particular model and find out how much the materials will cost. Don't forget to add the cost of your labour! Decide on your hourly rate and multiply it by the estimated number of hours you would spend on the model. Your selling price should include your total cost, plus some profit.

Pricing

When you buy a stamp collection for £10.00 and sell it for £11.00, are you really making money? Perhaps not: The price may not be enough to cover the cost of operating your business. This section will help you to understand the pricing of goods for sale.

Fact box ▶

Unit cost: The amount the seller pays for one item.

Mark-up: The amount added to the unit cost to find the **selling price**. Mark-up is usually a **percentage** of the **unit cost**.

The selling price of a jacket with a unit cost of £9.80 and a 25% mark-up is calculated this way.

Mark-up = 25% of £9.80
= 0.25 × £9.80
= £2.45

Selling price = Cost of item + Mark-up
= £9.80 + £2.45
= £12.25

Read the facts carefully and answer the questions.

1. In order to pay for operating costs and to make some net profit, the Candlelight Shop must sell candles with a 20% mark-up on cost. What should be the selling price for these candles?

2. Lower prices often create more sales. If your prices are higher than those in other shops, you may not be able to sell your goods. The following chart shows how the lower mark-up affected the total sales of a calculator. Find the missing mark-ups and totals.

Cost	% Mark-up	£ Mark-up	Total number of calculators sold	Total mark-up or gross profit
£12.00	20%	*£2.40*	200	*£480.00*
12.00	25%		175	
12.00	30%		100	
12.00	35%		50	

Which mark-up percentage had the highest gross profit?

Type of candle	Cost	20% Mark-up (on cost)	Selling price
Mushroom	£0.85	*0.20 × 0.85 = 0.17*	*0.85 + 0.17 = £1.02*
Egg	£0.60		
Animal	0.95		
Cartoon character	£1.05		

3. Find the selling price for each fish tank based on the cost of materials and percentage of mark-up.

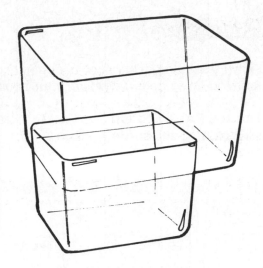

Size	Cost of materials	% Mark-up	£ Mark-up	Selling price
Small	£13.50	250%	£33.75	£47.25
Medium	£14.25	300%		
Large	£15.70	290%		
Extra large	£17.00	310%		

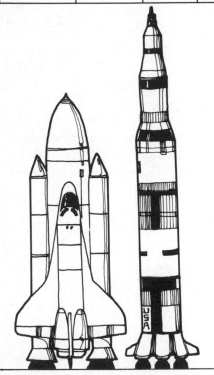

4. Suppose you build model spaceships and want to sell them at a profit. The materials for one model cost £4.50. To pay for your labour and other expenses, you must price your models with 400% mark-up on cost. What is the selling price of one model spaceship?

Mark-up = 400% × £4.50

= 4 × ....

= ....

Selling Price = £............. + £.............

= £.............

On your Own

When you have a lot of things to sell, you can easily forget how much each one cost you. One way to remember without showing it to buyers is by using a secret code. Here is an example:

```
P R O F I T A B L E
1 2 3 4 5 6 7 8 9 0
```

The secret code for an item that costs £6.31 is T.OP. What code will be used for each of these costs – £7.76, £9.24, £0.24, £67.54?

Make up a code word of your own.

Bookkeeping

How's business? Your answer depends on what your records show. This section is about keeping up-to-date records of your business activities.

This is a CASH RECORD. Fill in each missing balance by **adding amounts received and subtracting amounts paid out.**

Cash Record

DATE		EXPLANATION	RECEIVED		PAID OUT		BALANCE	
Nov	1	Balance brought forward	545	60			545	60
Nov	5	New cardigans			210	00		
Nov	9	Sales	302	95				
Nov	14	Paper supplies			22	80		
Nov	15	Express Real Estate Co.			250	00		
Nov	16	Sales	550	00				
Nov	23	Sales	620	50				

This is a SALES REPORT. You use it to find total sales. It also shows you which items sell the most and the least. Fill in the missing pound totals per week in the Total column and the totals per item along the bottom.

Sales Report

	DEPARTMENTS			TOTAL
	KNITWEAR	SHIRTS	BEACHWEAR	
November 9	200.95	61.50	40.50	
November 16	325.25	170.50	54.25	
November 23	237.80	180.30	202.40	
November 30	420.00	215.60	208.70	
TOTAL				

This is a RECORD OF PURCHASES. It shows the date, payee (person or company paid), and the amount paid for purchased goods. This record is helpful when you want to know the cost of stock. Fill in the missing totals along the bottom.

Record of Purchases

DATE		PAYEE	AMOUNT PAID		KNITWEAR		SHIRTS		BEACHWEAR		
Nov	5	Fashion Gear Ltd	210	00	210	00					
Nov	26	Sunshine Bikinis	195	20					195	20	
Nov	27	Mayfair Shirt Co.	88	95			88	95			
Nov	28	Golden Sands Beachwear	54	25					54	25	
Nov	29	Embassy Garment Co.	90	00			90	00			
Nov	30	Briggs & Sons	77	50	77	50					
		TOTAL									

Report of Operating Expenses

DATE		PAYEE	AMOUNT PAID		ADS		PHONE ELECTRICITY RENT		SUPPLIES		OTHER	
Nov	14	Paper Bag Co.	22	80					22	80		
Nov	15	Express Estate Agent	250	00			250	00				
Nov	26	Daily Advertiser	60	00	60	00						
Nov	26	Telecom	75	40								
Nov	27	Ajax Transport	25	50								
Nov	29	Advertisers Ltd	50	00								
		TOTAL										

This is a REPORT OF OPERATING EXPENSES. It is a detailed picture of how much it costs you to run your business. The report includes the date, payee, amount paid, and what you bought. Fill in the missing entries and totals.

On your Own

Use the records you have made to find out if this business is making or losing money.

Reading a Graph

How can you chart the ups and downs of your business? Use graphs. They are easy to read and interesting to look at. This section will help you to understand how to use some kinds of graphs.

A graph is a kind of picture. Here are different kinds of graphs. Each one answers a different question about operating expenses.

A. Line graph
Are your expenses going up or down, or can't you tell? In a line graph, a dot is placed where the vertical line meets the horizontal line. In February, the expenses were £700.

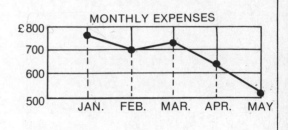

B. Bar (or column) graph
Which month was the most expensive or the least expensive? You can compare one month with another.

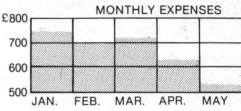

C. Circle graph

What portion of the budget do you spend on each item?

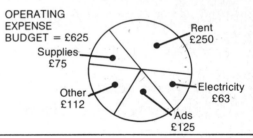

Complete this line graph to show monthly sales from August to December. Place dots on the graph showing these points:

a. Aug. – £50

b. Sept. – £100

c. Oct. – £200

d. Nov. – £175

e. Dec. – £250

Connect all the dots.

Are sales going up or down?   ....

Why do you think this is happening? ....

ACE SPORTS FOOTBALL EQUIPMENT
(Monthly Sales)

This bar graph compares the total sales of each item. Study the graph and answer the questions.

W = Shirts

X = Jeans

Y = Shorts

Z = Other

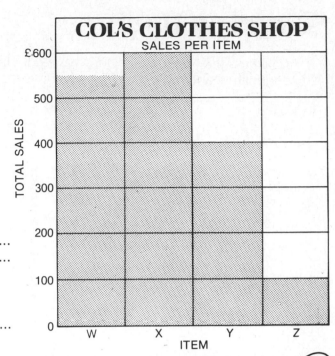

COL'S CLOTHES SHOP
SALES PER ITEM

a. Which item was the best seller? ....

b. Which item sold the least? ....

c. What was the difference between the sales of jeans and the sales of shorts? ....

d. Which items are closest in total sales? ....

e. What were the combined total sales of shirts and shorts? ....

You want to show what portion of your holiday income comes from each of your four jobs. In which part of your circle graph will you put these amounts?

LM (Lawnmowing)	= £25.00
NR (Newspaper Round)	= £18.75
BS (Babysitting)	= £50.00
CW (Car Washing)	= £6.25

Total Income ....

Write the letters and the amounts in the graph. What percentage of your income comes from:
1. LM ....
2. BS ....

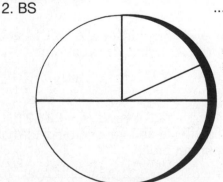

On your Own

These two graphs show the same information but give very different pictures. Why?

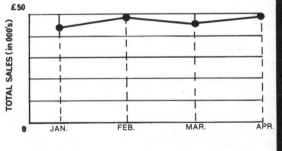

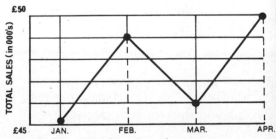

Find graphs used in advertisements. What is the advertiser trying to tell you!

Looking Back

1. Calculate the total hours, gross pay, tax, deductions and net pay.

Hr/day **7**	P.A.Y.E. tax: **30%**
Days/wk **6**	Social security **£9.42**
Total hours	Other deductions **£1.35**
	Total deductions
Rate **£2.30**	
Gross pay	Net pay

2. Stella works two days each week. First calculate her total hours in the office for one week. Next, subtract the lunch breaks. Find the total hours she might work in four weeks. Then calculate her average time per day.

Tuesday	5hr 45 min
Thursday	+ 8hr 30 min
1 week total	hr min
Breaks	− 1 hr 45 min
Actual time	hr min
	× 4 weeks
Total in 4 weeks	hr min

Average time per day = Total in 4 weeks
                                    Number of days worked in 4 wks

What is Stella'a average?

.... hr ....min

3. Jim earns 2.5p for each newspaper he delivers. How much does he make after delivering sixty newspapers?

....

Brian earns 4% commission for each car he sells. If he sells a car for £4,200, how much commission does he earn?

....

4.

Week Ending   4 April
Name   Michael Tan

DAYS	IN	OUT	IN	OUT	Daily Total	Ord. hours	Overtime (T½)	(DT)
M.	8–00	12–00	1–00	5–00	**8**	**8**		
Tu.	7–30	12–00	1–00	5–30	**9**	**8**		
W.	7–00	12–00	1–00	6–00	**10**	**8**		
Th.	7–00	12–00	12–45	6–45	**11**	**8**		
Fri.	7–15	12–30	1–15	4–00	**8**	**8**		
Sat.								

WEEKLY TOTAL

TOTAL ORD. HOURS   _____

(T½) ..... hr x 1.5   _____

(DT) ..... hr x 2   _____

TOTAL EQUIVALENT HOURS _____

RATE   **£2.20**

GROSS EARNINGS   _____

Fill in this time card with the following:

a. Overtime hours (T½ = time and a half; DT = double time for more than 2 hours overtime)

b. Weekly total hours

c. Total equivalent hours

d. Gross earnings

5. Suppose you bought a radio for £14 and sold it for £23. What was your gross profit?

....

In selling the radio, you spent £2.50 for advertisements and transportation. What was your net profit?

....

6. In order to pay for operating costs and to make a net profit, a photographer must take pictures with an 85% mark-up on cost. What should be the selling price of each size picture?

Size (cm)	Cost	85% Mark-up on Cost	Selling Price
6 by 6	£0.60		
12.5 by 18.0	1.20		
20 by 25	2.40		

7. Fill out this cash record with the following information:

1 May: the balance brought forward is £500.00   4 May: received £150 from sales
2 May: paid £250 rent   6 May: paid *Evening Post* £65 for advertisements

Date	Explanation	Received	Paid Out	Balance

8. Match each graph with the question it answers (A, B, or C).

A. How do you divide your budget?

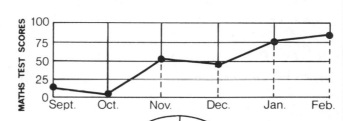

B. Who earned the most or the least money?

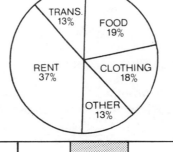

C. Are you improving your maths skills?

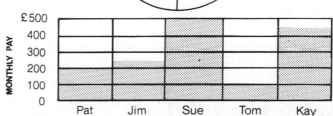

From the relevant graph answer the following questions:

a. What is your January maths test score? ....

b. How much money does Pat earn per month? ....

c. What percentage of your budget is spent on food? ....

Skills Survey

1. Add:

```
 13.50
  6.18
  3.10
  1.25      1 hr 15 min     7 hr 20 min
+ 0.60    + 3 hr 10 min   + 1 hr 55 min      27.90 + 11.69 + 6.60 + 2.90 + 0.90 = .........
```

2. Subtract:

```
£1575.40    £1113.00      3 hr 50 min     2 hr 10 min
-  342.20   -  435.67   - 1 hr 45 min   - 1 hr 25 min      £198.50 - £52.92 = .........
```

3. Multiply:

```
£15.50    3.50 × 0.25 = .........    3 hr 10 min     4 hr 35 min
 × 35     4.20 × 1.5  = .........       × 3             × 4
```

4. Divide:

```
70 )145.60    3 )9 hr 6 min    2 )3 hr 12 min    10.60 ÷ 530 = .........
```

5. Round to two decimal places:

0.278 ....   0.642 ....   £35.791 ....   684.085 ....
£17.998 ....   29.6387 ....   99.0084 ....   0.00634 ....

6. Calculate these percentages:

500% of £3.00 = ...   6% of £450.00 = ...   25% of £184 = ...
350% of £0.36 = ...   8% of £0.84 = ...   1.5% of £23.00 = ...

Branching Off

★ Interview one or two people who own small businesses. Ask them what kind of accounts they have to keep.

★ Find out the difference between wholesale price and retail price. How much discount do stores usually get from wholesalers?

UNIT 5

These days everyone wants to save money. This unit will show you some ways to do it.

Energy and Money Savers

Save a Watt

You might sometimes wonder why your electricity bills get bigger and bigger. You cannot always blame the electricity board. Sometimes the cause may be you and your appliances. This section will help you understand the cost of electricity used in your home.

Fact box

★ A **watt** is the unit for measuring electric power.

★ 1 kilowatt = 1000 watts.

   kW.h means kilowatt hour.

★ **1 kilowatt hour** = 1000 watts of electricity used in one hour. For example, a 100 watt light bulb left on for 10 hours uses 1 kilowatt hour of electricity (100 × 10 = 1000).

★ To calculate the yearly cost of electricity used by an appliance, multiply the estimated kW.h used each year by the cost of each kW.h.

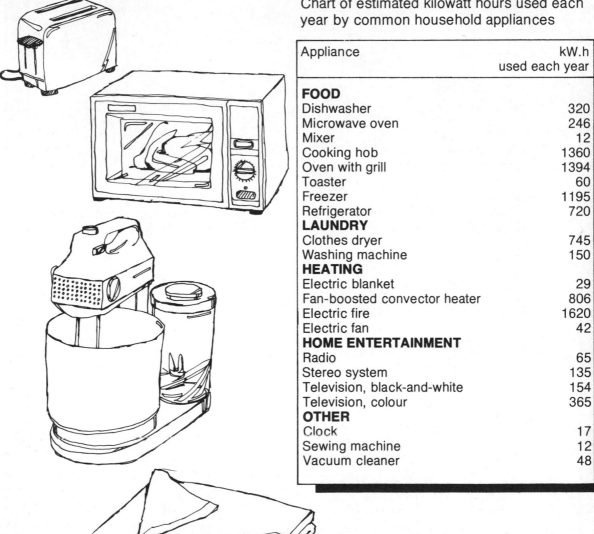

Chart of estimated kilowatt hours used each year by common household appliances

Appliance	kW.h used each year
FOOD	
Dishwasher	320
Microwave oven	246
Mixer	12
Cooking hob	1360
Oven with grill	1394
Toaster	60
Freezer	1195
Refrigerator	720
LAUNDRY	
Clothes dryer	745
Washing machine	150
HEATING	
Electric blanket	29
Fan-boosted convector heater	806
Electric fire	1620
Electric fan	42
HOME ENTERTAINMENT	
Radio	65
Stereo system	135
Television, black-and-white	154
Television, colour	365
OTHER	
Clock	17
Sewing machine	12
Vacuum cleaner	48

Use the figures in the chart to estimate the yearly cost of electricity in these exercises:

1.

House of the Evans Family		
Appliances	kW.h used each year	Yearly cost at 5p per kW.h
Cooking Hob	1360	1360 x 5p = £68·00
Toaster		
Refrigerator		
Electric fan		
Radio		
TV (B & W)		
Clock		
Sewing machine		
TOTAL		

4.

House of the Jackson Family		
Appliances	kW.h used each year	Yearly cost at 5p per kW.h
Microwave oven		
Dishwasher		
Refrigerator		
Colour TV		
Stereo		
Electric fan		
Vacuum cleaner		
TOTAL		

2. Suppose the Evans decide to replace their black-and-white television with a colour one, how much more will they pay for electricity each year?

....

3. The Evans decide to buy an automatic washing machine. How much will this add to the total cost of electricity per year?

....

5. The Jacksons plan to buy a freezer, a mixer, and a clothes' dryer. How many more kW.h will they use each year?

....

6. What is the total cost of the Jacksons' yearly electricity including the appliances in Exs. 4 and 5?

....

On Your Own

List some of the appliances in your home. Find out an average cost for 1 kW.h from your local electricity board. Calculate the yearly cost of each appliance based on the chart on the opposite page.

Get the Most from Your Car

World fuel shortages keep pushing up the price of petrol. Some cars travel much farther than others on the same amount of petrol. This section will help you to understand the fuel costs for your car.

Fact box ▶

★ **m.p.g.** (miles per gallon) = number of miles travelled on one gallon of petrol.

★ One way to calculate your car's petrol consumption is to divide the number of miles travelled by the number of gallons of petrol used to travel this distance.

You travel 270 miles and use 9 gallons of petrol.

$$\text{Petrol consumption} = \frac{270}{9}$$
$$= 30 \text{ m.p.g.}$$

★ To check your petrol consumption, start with a full tank. Write down the miles showing. Next time you fill up, write the new mile-reading and the gallons from the pump. Subtract the old reading from the new and divide by the number of gallons shown on the pump.

1. Calculate the miles each car has done between "fill-ups" and complete the chart.

	Ford Escort	Mini Metro	Rover 3500	Datsun Sunny
(N) New Reading (m)	6420	3246	14110	18760
(O) Old reading (m)	6148	2976	13845	18375
m travelled (subtract O from N)	272			

2. The Ford Escort travelled 272 miles and used 8 gallons of petrol. What was its petrol consumption?

Petrol consumption = $\frac{272}{8}$

= ....m.p.g.

3. The Rover 3500 takes 11 gallons to fill up. How many miles will it travel on one gallon of petrol?

.... m.p.g.

The figures in the chart give the government test figures for petrol consumption for three types of driving situation. 1. Town driving 2. Cruising (constant 56 m.p.h.) 3. Fast cruising (constant 75 m.p.h. – but don't try doing it!)

4. Four examiners take out different cars for a test drive. They each have 9 gallons of petrol and are instructed to use 3 for "town driving", 3 for "cruising" and 3 for "fast cruising". How far does each car travel on the petrol?

a. Ford Cortina 1.3L

Town driving	Cruising	Fast cruising	Total miles
$28.0 \times 3 = 84.0$	$37.7 \times 3 = 113.1$	$27.7 \times 3 = 83.1$	$84.0 + 113.1 + 83.1 = 280.2$ miles

b. Mini Metro 1.3S

Town driving	Cruising	Fast cruising	Total miles

c. BMW 732i

Town driving	Cruising	Fast cruising	Total miles

d. Renault 5TL

Town driving	Cruising	Fast cruising	Total miles

PETROL CONSUMPTION CHART			
Government Test Figures		Miles per gallon	
Make of car	Town driving	Cruising	Fast cruising
Ford Cortina 1.3L	28.0	37.7	27.7
Mini Metro 1.3S	32.8	51.2	37.9
Opel Kadett 13S	28.8	47.1	34.0
Triumph TR7	24.1	40.5	31.5
Ford Escort 1.1HC	34.9	49.6	36.2
Renault 5TL	44.8	57.6	41.5
BMW 732i	15.1	33.2	26.2
Toyota Corolla 1300	29.8	42.1	25.8
Volkswagen Golf 1.3L	26.9	42.8	31.7
Datsun Cherry 1.0	32.8	44.8	30.7
Talbot Horizon GL 1.3	31.7	44.1	31.4
Fiat Strada 75	26.4	43.7	32.9

On Your Own

Ask a mechanic what difference the following would make to petrol consumption:

a. An 8-cylinder engine, rather than a 6-cylinder engine;

b. Automatic transmission;

c. Power brakes, power steering, electric windows;

d. Five-speed gearbox.

Save Money, Save Lives

VRR-ROOM! So you like fast starts and high speeds! Here's news for you – you're throwing money away. Your car may have been advertised with a fuel consumption of 35 miles per gallon (m.p.g.). Your driving habits can easily lower your mileage to 25 m.p.g. This section is about improving driving habits and calculating savings on petrol.

Fact box →

Petrol-saving habits

★ Avoid fast starts and stops.

★ Drive at reduced speeds. When parked, turn off the engine. (A car idling for six minutes uses as much petrol as driving 1 mile at 30 m.p.h.)

★ Keep your car in good running condition. Have it serviced regularly.

★ By driving sensibly and keeping your car in good shape, you can save at least 15% on petrol costs – and save lives.

The bar graph shows you that for the same distance travelled, the faster you drive, the more petrol you use. Use the graph to answer Exs. 1-6.

Petrol used at different speeds
for the same distance (400 miles)

GALLONS

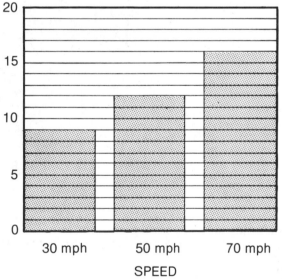

1. To cover the 400 miles, you use 9 gallons of petrol at 30 m.p.h. How many gallons would you use at 50 m.p.h?

2. How many gallons would you use if you drove the 400 miles at 70 m.p.h?

3. How much petrol could you save by driving at 50 m.p.h. instead of 70 m.p.h?

4. How much petrol could you save by driving at 30 m.p.h. instead of 70 m.p.h?

5. If petrol cost £1.60 a gallon, calculate the cost of making the trip at the different speeds. Fill in the chart.

m.p.h.	Gallons used	Total cost of petrol
30		
50		
70		

6. How much money could you save by doing the trip at 50 m.p.h. instead of 70 m.p.h?

....

Use the Fact Box to answer Ex. 7.

7. On a trip from London to Dover, Peter Beckley took the turn off to Maidstone instead of taking the bypass round it. Because of heavy traffic, his car was stopped, with the motor running, for at least twelve minutes. Peter wasted as much petrol as driving

.... miles at .... m.p.h.

On Your Own

Choose a car that you would like to buy. Find out its petrol consumption. Plan a trip and estimate what it would cost you in fuel.

8. By driving carefully, each of these drivers can get 15% extra miles per year, without extra petrol costs. Calculate the number of extra miles.

Driver	Miles driven in one year	Extra miles possible
Tony Russell	8500	*1275*
Sue Marshall	7420	
Dilip Khartri	14320	
Pam Smith	12940	

15% of
8500
= 0.15 × 8500
= 1275

Do It Yourself

More and more people are decorating their homes and doing minor repairs themselves. Want to join them? This lesson will help you understand how to measure area in your home so that you can decorate it yourself.

Fact box

For a room this shape

Area = Length × Width

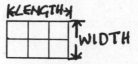

Area is measured in square metres (m²) or square centimetres (cm²).

If the length is 3m and the width is 2m:
Area = 3m × 2m = (3 × 2)m² = 6m²

To change centimetres to metres: divide centimetres by 100:
200 cm = (200 ÷ 100)m = 2m

To change metres to centimetres: multiply by 100:
4m = (4 × 100)cm = 400cm

Use the Fact Box and the advertisement to solve the problems in Exs 1-3.

ATCOST CARPET WAREHOUSE		
CARPET	COST per SQUARE METRE	COST INC. LAYING/SQ.M.
Lightweight Cord	£3.95	£4.40
Heavy Duty Cord	£4.40	£5.10
Twist Pile	£6.95	£7.95
All Wool Axminster	£8.80	£10.05

1. Write in the area in square metres for each room in this floor plan:

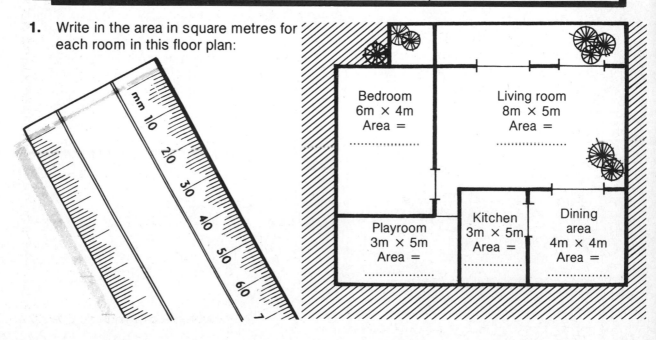

Bedroom
6m × 4m
Area =
....

Living room
8m × 5m
Area =
....

Playroom
3m × 5m
Area =
....

Kitchen
3m × 5m
Area =
....

Dining area
4m × 4m
Area =
....

2. Fill in this chart to calculate how much it costs to buy carpet for each room and how much you can save by laying it yourself. The playroom is done for you.

	Living room	Bedroom	Dining area	Playroom	
Carpet Type	Axminster	Twist Pile	Heavy duty cord	Lightweight cord	
Length in m				5 m	
Width in m				3 m	
Area in m²				15 m²	← 5 m × 3 m
Cost per m² (inc. laying)				£4.40	
Total cost (inc. laying)				£66.00	← 15 × £4.40
Cost per m² (without laying)				£3.95	
Total cost (without laying)				£59.25	← 15 × £3.95
Total cost for carpet and laying				£66.00	
Total cost for carpet only				£59.25	← subtract
Savings				£6.75	

3. Suppose you decide to lay ceramic tiles in the kitchen in Ex. 1. The ceramic tiles that you like come in three sizes. How many tiles will you need in each size? The first one is done for you.

Kitchen floor area = 3 m × 4 m

$\qquad\qquad\qquad$ = 300 cm × 400 cm

Kitchen floor area = 120 000 cm²

Divide this area by tile area.

Ceramic Tile Size	Tile Area	Number of Tiles Needed
10 cm × 5 cm	50 cm²	120 000 ÷ 50 = 2400
15 cm × 10 cm	....	120 000 ÷ .... = ...
20 cm × 15 cm	....	120 000 ÷ .... = ...

On Your Own

Draw the shape of your bedroom and kitchen. Fill in the measurements and calculate the area of each room in square metres.

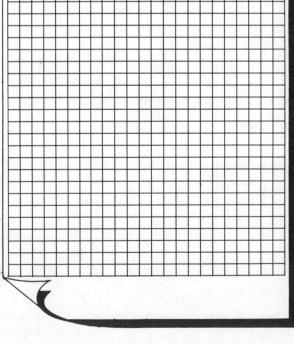

End-Of-Season Sale!

You hear about the end-of-season sale and come prepared with a shopping list. But how much will you really save on each item? Can you tell from the advertisements? This section will help you find the amount of discount and new sale price, from given percentages.

★ To calculate the amount of discount: multiply the percentage (in decimal form) by the original selling price.

The jacket was selling for £14.25. It now has a discount of 20%:
20% of £14.25 = 0.20 × £14.25
= £2.85

★ To calculate the new discounted price: subtract the amount of discount from the original selling price.

The jacket was selling for £14.25. The discount is £2.85.
New discounted price = £14.25 − £2.85
= £11.40

50% off sportsware
20% off items marked with ✦
15% off all other items

1. Use the advertisement to determine the percentage discount for each item listed. Then calculate the amount of discount and the discounted price.

Item	Price	% Discount	Amount of discount	Discounted price
Swimsuit	£8.60	50	£4.30	£4.30
Jacket ✿	£14.25	20	£2.85	£11.40
Dress	£12.80			
Tracksuit	£18.78			
Sandals	£10.40			
Cricket sweater	£9.88			
Raincoat ✿	£19.30			
Tennis Dress	£10.74			
Boots	£23.60			
Coat	£32.60			
TOTAL			TOTAL	

2. Three people bought the same type of coat in different shops. Find the discounted price of the coat in each shop.

Shop	Price Tag	% Discount	Amount of Discount	Sale Price
Jacki's	£44.95	20%	£..................	£..................
Top Coats	£55.60	25%	£..................	£..................
In-Gear	£40.80	15%	£..................	£..................

In which shop was the coat the cheapest? ....

On Your Own

Look for sales advertised in the newspaper. List the things you would like to buy and calculate the discounted price from the advertised percentages.

Item	Sale Price
....	....
....	....
....	....
....	....

Buy More, Pay Less?

How can you pay less when you buy more? This section will help you find out how you often save money by buying things in large quantities.

* **Unit price:** The amount you pay for one item. To calculate the unit price of one item in a package: divide the package price by the number of items in the package.

  Example: What is the price of each tin of fruit if a carton of twenty-four tins costs £9.60?

  ```
           0.40
  Unit price = 24 9.60   Unit price = £0.40
              0
             ──
             96
             96
             ──
             00
  ```

* To find the total cost of the same number of tins bought separately at £0.46 each:

  Total cost of twenty-four tins = £0.46 × 24
  = £11.04

* To find the amount saved by buying the carton, instead of separate tins:

  Total cost of items
  bought separately = £11.04
  Carton price = £9.60 −
  Amount saved = £1.44

Read the facts in Exs. 1-5 carefully. Decide whether you pay less by buying more.

1. One bar of soap costs 29p. A pack of four costs £1.10. How much do you save by buying the pack of four instead of four separate bars of soap?

   ....

2. A 500 ml bottle of shampoo costs £1.56. A 250 ml bottle costs 84p. How much do you save by buying the larger bottle instead of two small bottles?

   ....

3. A 10 kg bag of sugar costs £2.49. You can buy smaller bags in the following amounts:

   1 kg bag £0.29

   2 kg bag £0.55

   5 kg bag £1.32

   How much would 10 kilograms of sugar cost if you bought it in:

   a. 1 kg bags? ....

   b. 2 kg bags? ....

   c. 5 kg bags? ....

   How much do you save by buying the 10 kg bag instead of:

   d. 1 kg bags? ....

   e. 2 kg bags? ....

   f. 5 kg bags? ....

4. A two litre tin of orange juice costs 87p. A litre carton costs 43p. Do you save money by buying the larger container?

   ....

5. Read this advertisement.

PiCNIC SET !
only £ 14.99

If bought separately the items in the set would cost:

Cool box £4.50

Vacuum flask £3.75

6 plastic plates £3.00

6 plastic cups £2.40

6 sets picnic cutlery £1.98

Barbecue set £4.99

a. What is the total cost of the items bought separately?

....

b. How much do you save by buying the set instead of buying the items separately?

....

c. Suppose you don't need the picnic cutlery. Is it still cheaper to buy the set?

....

d. You only need a cool box, a vacuum flask and a barbecue set. Should you buy the whole set or buy the items separately?

....

On Your Own

Find items in a supermarket which are available in various sizes. Compare prices and check whether buying more costs you less. Start with the example given.

Item	Size 1	Cost	Size 2	Cost	Better value
Coffee	100 g jar	80p	750 g tin	£5.12	

Looking Back

1. Find the yearly cost of using each appliance

Appliance	kW.h used each year	Yearly cost at 5p per kW.h
Oven with grill	1394	
Refrigerator	720	
Colour TV	365	
Stereo system	135	
Sewing machine	12	
Electric fire	1620	
TOTAL COST		

2. How much would you save if you replaced the electric fire in Ex. 1 with a fan-boosted convector heater that uses 806 kW.h per year?

3. Three people driving different cars travel 60 miles under town driving conditions and 200 miles at a cruising speed of a constant 56 m.p.h. How many gallons of petrol are used in a day by each driver?

	Honda Civic	Ford Ghia	Jaguar
Town driving (m.p.g.) (miles per gallon)	30	25	15
Cruising speed (m.p.g.)	50	40	30
No. of glns. used in town			
No. of glns. used at cruising speed			
Total gallons used			

4. If petrol costs £1.90 a gallon, how much would the Honda driver in Ex. 3 spend on petrol for the day's journey?

5. Your car uses 9 gallons of petrol to travel 324 miles. How many miles per gallon is that?

6. Your car's average fuel consumption is 34 m.p.g. The petrol tank holds 8.5 gallons. How far could you travel on a full tank?

7. Find the area in square metres of each room in this floor plan

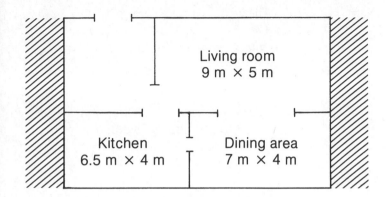

Living room: area = ...
Dining area: area = ...
Kitchen: area = ...

8. Three people bought the same type of suit in different stores. Find the discounted price of the suit in each store.

Store	Price Tag	% Discount	Amount of Discount	Sale Price
Jim's	£89.50	30%		
Pat's	£81.40	25%		
Len's	£74.95	20%		

9. Read the advertisement carefully. Then answer the questions.

a. What is the unit cost of 1 bottle in a case?

....

b. What is the unit cost of 1 bottle in a six-pack?

....

c. How much do you save by buying a case instead of buying twenty-four separate bottles?

....

FRESH ORANGE JUICE

1 CASE   £9.12
(twenty-four 1 litre bottles)

6-PACK   £2.40
(six 1 litre bottles)

1 litre bottle 46p

Skills Survey

1. Multiply the KW.h by £0.05 to find the monthly cost of electricity.

	B&W TV	Radio	Toaster
kW.h used each month	10	8	5
Cost of each kW.h	.05	.05	.05

2. Change to metres:

200 cm = ...

125 cm = ...

396 cm = ...

3. Change to centimetres:

3 m = ...

1.7 m = ...

10.25 m = ...

4. Calculate in square metres (m²)

7 m × 3 m = ...

12 m × 21 m = ...

10 m × 2.5 m = ...

5. What is the area of the following rooms?

a. Bedroom, 3 m × 3.5 m ....

b. Hall, 1.5 m × 2.6 m ....

c. Shower, 1 m × 0.75 m ....

6. Find the percentages. Round your answer to the nearest penny.

25% of £45.37 = ...

35% of £79.50 = ...

20% of £185.00 = ...

7. What is the price of one bottle in a pack of six bottles for £3.78?

....

8. Which is cheaper, two for 99p or one for 50p?

....

9. A two-litre tin of fruit costs £2.79 and a one-litre tin costs £1.43. How much money do you save by buying the bigger tin?

....

Branching Off

★ Boxes are measured in terms of volume, expressed in cubic centimetres and cubic metres. Measure your wardrobe at home and find its volume. Make a list of the other things at home that you could measure by volume.

★ You have learned how to measure the area of a square and a rectangle. How do you measure the area of other geometric figures such as triangles and circles?

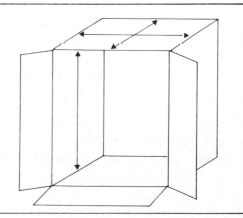

UNIT 6

Your maths skills can come in handy when you're having fun!

Leisure Maths

Where Does Your Team Stand?

The sports pages of our daily newspapers are filled with statistics, numbers of wins and defeats, records broken, league tables. You can calculate the statistics for your favourite team. These pages will help you understand how to work out team positions in the first division of the Football League.

The table shows the number of games played, won and drawn by the teams in the first division mid-way through the 1983/84 season. They are in alphabetical order. Calculate the number of points scored by each team and rearrange them in their correct league positions.

	P	W	D	Pts.
Arsenal	29	11	5	38
Aston Villa	28	11	8	......
Birmingham	28	9	6	......
Coventry	28	10	9	......
Everton	27	9	9	......
Ipswich	28	9	5	......
Leicester	28	8	8	......
Liverpool	29	17	4	......
Luton	28	12	4	......
Manchester United	29	15	10	......
Norwich	29	11	9	......
Notts. County	28	5	6	......
Nottingham Forest	29	16	5	......
Q.P.R.	28	14	4	......
Southampton	28	14	6	......
Stoke	29	7	8	......
Sunderland	28	8	9	......
Tottenham	29	11	8	......
Watford	29	12	5	......
West Bromwich	28	9	5	......
West Ham	29	15	5	......
Wolverhampton	28	4	7	......

1. From the table how many games did the following teams lose?
   a) Arsenal b) Luton c) Stoke
   d) Liverpool e) Watford f) Coventry.

★ In Rugby Union the team with the highest percentage of points heads the "Order of Merit" table. Win = 2 points, Draw = 1 point, Defeat = 0 points.

★ To calculate the points percentage follow the steps in the example.

Example: Cardiff played 6, won 4, drawn 1, lost 1.

Divide the number of points won by the total number of points possible. Multiply the answer by 100 to give the percentage.

Points won = 9 (that is, $4 \times 2 + 1 \times 1$)
Points poss. = 12 (that is, 6×2)
Pct = $9 \div 12 \times 100$
  = 0.75×100
  = 75

$$
\begin{array}{r}
0.75 \\
12\overline{)9} \\
0 \\
\overline{90} \\
84 \\
\overline{60} \\
60 \\
\overline{0}
\end{array}
$$

Use the RFU Merit Table to answer the following questions:

1. Which team has the highest points percentage? ....

2. Which team is highest on the merit table: Rosslyn Park, Wasps or Harlequins? ....

3. Which team has the lower points percentage: Richmond or London Irish? ....

4. Which team has the lowest points percentage overall? ....

5. Place the teams in the merit table in order by putting the best percentage record first.

   a. ....

   b. ....

   c. ....

   d. ....

   e. ....

   f. ....

   g. ....

   h. ....

   i. ....

   j. ....

RFU Merit Table						
	P	W	D	L	Pts.	Pct.
Wasps	8	4	0	4	......	......
London Irish	10	7	1	2	......	......
Saracens	6	2	0	4	......	......
London Welsh	5	4	0	1	......	......
Met. Police	6	1	1	4	......	......
Richmond	5	3	1	1	......	......
Harlequins	5	2	0	3	......	......
Rosslyn Park	10	2	3	5	......	......
London Scottish	6	4	0	2	......	......
Blackheath	5	1	0	4	......	......

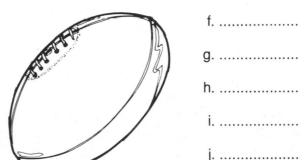

On Your Own

Find out how the points system works in the County Cricket Championship.

What's The Score?

When you watch cricket on television, do you ask yourself, "How do they get that score?" In this section you will learn how to keep the score and how to calculate batting averages.

Fact box

Scoring

Batsmen in cricket can score runs in the following ways:

A **four:**   Hitting the ball so that it crosses the boundary line, but hits the ground inside the boundary line first.

A **six:**   Hitting the ball so that it crosses the boundary line "on the full" (without first bouncing inside the boundary).

Runs can also be scored when a batsman hits the ball and runs the length of the pitch before a fieldsman can return the ball to the stumps. The number of runs depends on how many times the batsman runs the length of the pitch.

Batting Averages

n.o. = not out

A batting average is calculated by adding together a batsman's scores and dividing by the number of times he was dismissed (out).

Example: Geoff Boycott: 72, 48, 53 n.o., 146 n.o.
17
Total runs: 336
No. of times dismissed:
3 (5 minus 2 n.o.)
Batting average = 336 ÷ 3
= 112

1. This diagram shows how Ian Botham scored his runs in one innings. Fill in the spaces and calculate his score.

	Number	Runs
Fours	*5*	*20*
Sixes		
Other runs		
TOTAL		

- - - - - - means "on the full".

(See Fact Box)

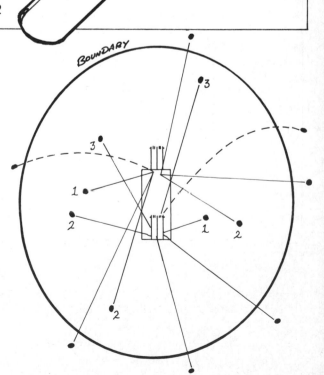

2. England v West Indies – One-day match

A score sheet records the runs in the order in which they were scored.
Calculate each batsman's score and the team totals. The first one is done for you.

ENGLAND	SCORE	TOTAL	WEST INDIES	SCORE	TOTAL
Brearley	2.4.3.1.1.4.4.2.2.3.1.2. 4.2.2.3.1.1.6.4.2.3.2.1	60	Rowe	4.2.1.2	9
Boycott	4.1.1.2.2.4.4.3.3.2.4.4 1.1.3.3.2.4.4.4.2.1.3.1.1.3		Greenidge	4.4.1.1.2.4.4.	
Randall	4.2.1.2.4.2.		Hayes	1.1.4.4.2	
Gooch	3.2.1.2.4.4.2.6.4.1.1.		Richards	4.3.1.4.4.2.4.3.6.4.4.2 1.2.2.6.4.1.6.4.2.1.1.4 4.4.4.2.3.4.6.3.1.4.4	
Gower	2.1.4.2.4		Kallicharran	4.	
Botham	4.4.2		Lloyd	2.4.4.1.1.1.3.6.4.4.2.4 4.4.1.3.3.2.6.4.4.1.6.2.	
Gatting	1.3.4		Murray	1.1.2.1.	
Edmonds	2.		Roberts	4.2.4.1.4	
Old	1.4.2.1.		Garner	2.1.	
Taylor	2.1.1.4.3		Holding	1.	
Hendrick	4.6.2.1.1.4		Croft	2.	
Extras		21	Extras		5
	TEAM TOTAL			TEAM TOTAL	

Which team won the match? .... By how many runs? ....

3. Calculate each batsman's batting average (see Fact Box).

	INNGS.	N.O.	RUNS	AVE.
BOYCOTT (34, 126, 53 n.o., 28, 39)	5	1	280	70.0
GOWER (6, 74, 50, 42, 18, 20 n.o.)	......	......	......	......
BOTHAM (40, 2, 68, 85, 0)	......	......	......	......
GOOCH (110, 8, 24 n.o., 34, 37 n.o.)	......	......	......	......
GATTING (42, 68, 4, 23, 36, 19)	......	......	......	......

Which batsman scored most runs?

....

Which batsman had the highest average?

....

On Your Own

Different sports have different systems of scoring. Find out how to score a game of tennis, squash, hockey, snooker, badminton and Rugby Union.

On The Move

How far are you going? How fast? How much time do you need to get there? These are some of the questions this section will help you answer.

Fact box ▶

How far?

Distance (in miles) = average speed × time
(average speed in m.p.h. multiplied by time in hours)

You travel at 40 m.p.h. for 4 hours
Distance travelled = (40 × 4) miles
= 160 miles

How fast?

Average speed (in m.p.h.) = distance ÷ time
(distance in miles divided by time in hours)

You travel 250 miles in 5 hours.
Average speed = (250 ÷ 5) m.p.h.
= 50 m.p.h.

How much time?

Time (in hours) = distance ÷ average speed
(distance in miles divided by average speed in m.p.h.)

You travel 150 miles at 60 m.p.h.
Time = (150 ÷ 60) hours
= 2.5 hours

Parts of an hour :

0.25 = ¼ of an hour = 15 minutes
0.5 = ½ of an hour = 30 minutes
0.75 = ¾ of an hour = 45 minutes

In these exercises, round your time to the nearest quarter of an hour.

You are on a motoring holiday in Wales. Use the FACT BOX and the road map to answer the questions in Exs. 1-5.

1. You start in Swansea. You decide to visit Monmouth and stay overnight. There are two main routes you can take. You can travel through Merthyr Tydfil or pass through Cardiff and Newport. From the map, which way is longer?

    ....

2. From Monmouth you drive to Carmarthen via Brecon. What is the distance travelled?

    ....

    What was your average speed if it took 3 hours to get there?

    ....

3. Staying overnight in Carmarthen, you decide next day to go on a day trip to Pembroke. On this journey you are able to average 30 m.p.h. for the whole journey, there and back. About how long does the whole journey take you?

    ....

4. The journey from Carmarthen to Dolgellau takes 6 hours at an average speed of 25 m.p.h. What is the distance from Carmarthen to Dolgellau?

    ....

5. From Dolgellau you travel via Wrexham, to spend the rest of your holiday at Colwyn Bay. If it takes you 3 hours to complete this journey, what is your average speed?

    ....

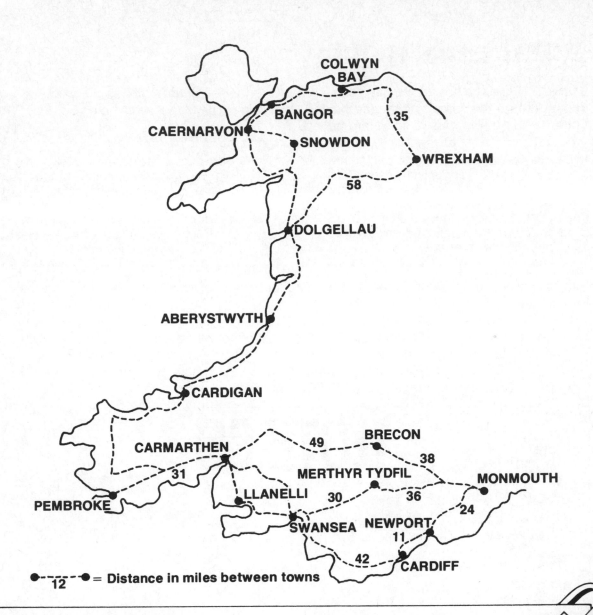

COLWYN BAY

BANGOR

CAERNARVON

SNOWDON

35

WREXHAM

58

DOLGELLAU

ABERYSTWYTH

CARDIGAN

BRECON

CARMARTHEN 49

38

MERTHYR TYDFIL

MONMOUTH

31

LLANELLI 30 36

24

PEMBROKE

SWANSEA NEWPORT

11

42 CARDIFF

●--~--● = **Distance in miles between towns**
   12

On your Own

Fill in the chart with the missing distance, travelling time, or average speed.

From Dolgellau to	Distance	Travelling Time	Average Speed
Cardigan	75 miles	3 hours	....
Pembroke	120 miles	....	30 m.p.h.
Newport	....	7 hours	36 m.p.h.
Llanelli	180 miles	....	40 m.p.h.

A Package Holiday

Everybody likes to have a summer holiday. The number of different prices for what appears to be the same holiday can be very confusing. This section will help you to understand how various holidays are priced.

Package holiday: Includes the cost of travel to your holiday resort and the cost of your accommodation.

Half board: Includes the cost of accommodation, breakfast and evening meal.

Full board: Includes the cost of accommodation and all meals.

Self-catering: You pay for accommodation and make your own meal arrangements.

Use the advertisement from a holiday brochure for the Isle of Wight to work out the cost of the holidays in Exs. 1-6.

Travel by car. The cost of the car ferry must be added to the holiday cost.
Ferry prices for any length of car:
Off Peak ADD £14 to holiday costs, Summer ADD £20 to holiday costs.
Summer savers (travel mid-week) ADD £14 to holiday costs.

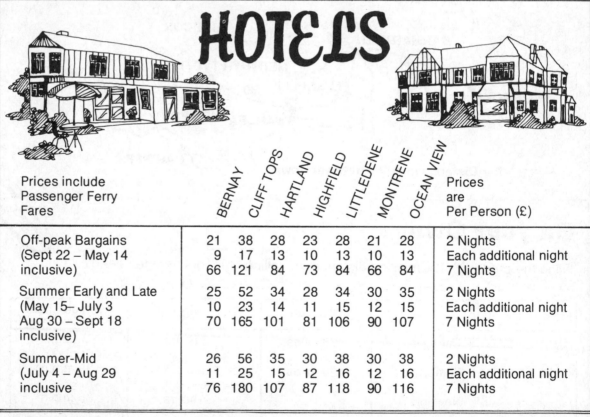

HOTELS

Prices include Passenger Ferry Fares	BERNAY	CLIFF TOPS	HARTLAND	HIGHFIELD	LITTLEDENE	MONTRENE	OCEAN VIEW	Prices are Per Person (£)
Off-peak Bargains (Sept 22 – May 14 inclusive)	21	38	28	23	28	21	28	2 Nights
	9	17	13	10	13	10	13	Each additional night
	66	121	84	73	84	66	84	7 Nights
Summer Early and Late (May 15– July 3 Aug 30 – Sept 18 inclusive)	25	52	34	28	34	30	35	2 Nights
	10	23	14	11	15	12	15	Each additional night
	70	165	101	81	106	90	107	7 Nights
Summer-Mid (July 4 – Aug 29 inclusive	26	56	35	30	38	30	38	2 Nights
	11	25	15	12	16	12	16	Each additional night
	76	180	107	87	118	90	116	7 Nights

Children's prices:
Off Peak – Aged 5-13 years half-price, Under 5 years FREE.
Summer – Aged 5-13 years two-thirds price, 5-9 years half-price, Under 5 years one-third price.

1. Mr and Mrs Green and their two children, aged 7 and 4, spent eight days at the Hartland Hotel from 16 May to 23 May. They travelled mid-week. How much did this cost them?

Cost for 7 nights = £101 (early summer prices)

1 addit. night	=	14 +
		£115

For 2 persons 2 × 115 = £230 ......A.

For child aged 7 = £50.50 (half-price)

1 addit. night	=	7.00 +	
		£57.50	......B.

For child aged 4 = £33.67 (one-third price to nearest penny)

1 addit. night	=	4.67 +	
		£38.34	......C.

Cost of car ferry = £14.00 ......D.

Total cost = A. + B. + C. + D.	£230.00
	57.50
	38.34
	14.00 +
Total cost =	£339.84

2. Frank and Sally Weston spend four days at the Ocean View Hotel in November. How much does this cost them, including the ferry?

....

3. Mr and Mrs Bartlett have 12-year-old twins. They take a week's holiday at the Cliff Tops Hotel in the middle of July, arriving on Saturday. How much does their holiday cost?

....

4. If the Bartlett family had stayed at the Littledene Hotel instead of the Cliff Tops, how much money would they have saved?

....

5. Ron and Mary Dean and their 4-year-old son Peter have a ten-day holiday staying at the Bernay Hotel in April. What is the cost of their holiday?

....

6. The Dean family enjoyed their holiday so much, they decide to spend another ten days at the Bernay Hotel in August. If they travel on Saturday, how much more expensive is their August holiday than their April holiday?

....

On Your Own

Most tour operators offer package holidays on the continent in summer and winter. Get some holiday brochures and choose a two-week summer holiday, and also a week's skiing holiday in the winter. Find the total cost for these holidays and compare them with similar holidays run by other tour operators.

A Camping Holiday

Camping has become a popular form of holiday. For the first-time camper, it is possible to hire all the equipment needed for a complete holiday.

Use the Hire Charge table and the Fact Box to answer questions 1-4

Equipment	Hire Charge A	B
Atlas Two-Man Ridge Tent	£0.95	£0.85
Explorer Motorist Tent (sleeps three)	1.30	1.10
Safari Four-Person Frame Tent	2.40	2.10
Safari Six-Person Frame Tent	2.80	2.45
Sleeping Bag	0.30	0.27
Camp Bed	0.20	0.17
Children's Folding Cot	0.26	0.23
Chair	0.20	0.17
Camping Table	0.26	0.23
Two-Burner Gas Stove	0.38	0.34
Insulated Cool Box	0.24	0.17
Standard Roof Rack	0.29	0.25
Gas Light	0.26	0.23

A = high season   B = low season

100

1. Peter and Jean Hayden are taking their son Sean and baby daughter Lucy on a 10-day camping holiday in July. What is the cost of hiring the equipment they need?

Equipment	No. reqd.	Cost/Day	No. of Days	Total Cost	
Safari four-person tent	1	£2.40	10	£24.00	1 × 2.40 × 10
Sleeping bags	3	0.30	10	9.00	3 × 0.30 × 10
Children's folding cot	1	0.26	10	2.60	1 × 0.26 × 10
Chairs	3	0.20	10	6.00	3 × 0.20 × 10
Table	1	0.26	10 ·	2.60	1 × 0.26 × 10
Two-burner gas stove	1	0.38	10	3.80	1 × 0.38 × 10
Total cost of hire of equipment = £48.00					

2. Mr and Mrs Reid decide to go on a camping holiday for eight days in May. They need to hire a six-person tent, two camp beds, four sleeping bags, four chairs and a table. Find the total hire cost.

Total . . . .

3. Ian and John Harper go on a two-week cycling holiday in August. They hire a two-man ridge tent, two sleeping bags, a gas light and a cool box. How much does it cost to hire this equipment?

Total . . . .

4. Mr Britain, a teacher, takes a group of ten pupils on a field trip to Wales for seven days in July. They travel in the school minibus, but need to hire the following equipment: one ridge tent, one four-person tent, one six-person tent, six sleeping bags, two cool boxes and a roof rack. What is the total hire cost?

Total . . . .

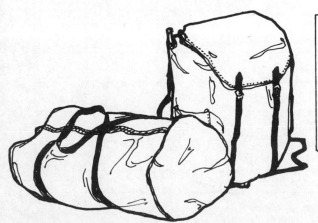

On your Own

You and three friends are going on a camping holiday. Make a list of essential equipment you would need.
Find out the cost of hiring the equipment from the hire charge list.

Foreign Money

You see advertisements in a foreign magazine. What do things really cost? This section will show you how to change one country's money (currency) into another.

To change English pounds into foreign currency X:

Multiply the amount in English pounds by the number of X in one English pound.

To change currency X into English pounds:

a. Divide the total amount of X by the number of X in one English pound.

b. Round your answer to the nearest penny. (English pound = pound sterling)

The exchange values given in this section may be used to do Exs. 1-7.
Currency values can change daily, so check with a local bank to find the day's rate if you want to send money overseas.

FOREIGN CURRENCY VALUES EQUIVALENT TO 1 POUND STERLING*	
AUSTRIA	26.50 Schillings
FRANCE	11.6 French francs
WEST GERMANY	3.8 Marks
ITALY	2320 Lire
SWITZERLAND	3.20 Swiss francs
SPAIN	220 Pesetas
UNITED STATES	1.48 Dollars
GREECE	145 Drachmas
* Values may change from day to day	

1. You are on a motoring holiday in Europe. When you arrive in each of the following countries, you change 50 pounds sterling (£50) into local currency. How much do you have in each currency?

FRANCE $50 \times 11.6 = 580$ francs

SWITZERLAND   .... Swiss francs

AUSTRIA   .... schillings

WEST GERMANY   .... marks

ITALY   .... lire

2. A holiday apartment in Spain costs £95 per week to rent. How many pesetas is that?   ....

3. A ski-lift pass cost John Dawson 44.80 Swiss francs. How many pounds is that?   ....

4. Peter saw a bike kit advertised in an American magazine for $66.60. What is the cost in pounds?   ....

5. A car in West Germany cost 21750 marks. What would it cost in English money?   ....

6. Your uncle in Greece sent you 2175 drachmas for your birthday. How much is that in pounds?   ....

7. On a day-trip to France you take £24 to spend in Boulogne. How much is that in francs?   ....

On Your Own

Make a list of the countries you would like to visit. Check with a bank how much the pound is worth in the currency of each country.

Looking Back

1. Fill in the **Pts.** (points) column and the **Pct.** (percentage) column and find the teams' position in the Merit Table. (Two pts. for a win, one pt. for a draw)

	P	W	D	L	Pts.	Pct.	Pos.
Bristol	8	5	0	3	.....	.....	.....
Gloucester	6	3	1	2	.....	.....	.....
Plymouth	4	1	0	3	.....	.....	.....
Camborne	5	1	0	4	.....	.....	
Exeter	5	2	0	3	.....	.....	.....
Bath	6	4	1	1	.....	.....	.....

2. What are the batting averages of these cricketers?

	Total	Batting Av.
Lloyd 27, 101, 64, 33	....	....
Richards 19, 54, 61 n.o., 70, 24 n.o.	....	....

3. How far would you go if you travel:

   a. At 50 mph for 4 hours?   ....

   b. At 37 mph for 5 hours?   ....

   c. At 42 mph for 2½ hours?   ....

   d. At 35 mph for 3½ hours?   ....

4. What is your average speed if you travel:

   a. 240 miles in 6 hours?   ....

   b. 185 miles in 5 hours?   ....

   c. 100 miles in 2½ hours?   ....

   d. 147 miles in 3½ hours?   ....

5. How much time would it take you to travel:

   a. 250 miles at an average speed of 50 mph?   ....

   b. 180 miles at an average speed of 45 mph?   ....

   c. 90 miles at an average speed of 60 mph?   ....

   d. 135 miles at an average speed of 30 mph?   ....

6. To stay at the Seaview Hotel the cost is £14 per night in the summer season and £10 per night at off peak times. Children are half price.

a. What is the cost for three people for four nights in the summer?

....

b. What is the cost for two adults and one child for five nights at an off peak time?

....

c. What is the cost for two adults and two children for three nights in the summer?

....

d. What is the cost for four adults and five children staying for five nights in April (off peak)?

....

Use this table to answer Exs.7-10

Holland	4.25 guilders	– 1 pound sterling
Japan	330 yen	– 1 pound sterling
Yugoslavia	196 dinar	– 1 pound sterling
(Remember these rates change from day to day.)		

7. You change £50 into the local currency of each of these countries. How much will it be in the local currency?

Holland ....

Japan ....

Yugoslavia ....

8. A Delft pottery plate sells for 21.25 guilders in Holland. How much is this in English pounds?

....

9. A carved wooden bowl costs 882 dinar in Yugoslavia. How much is this in English money?

....

10. A Sony television costs £260 in England. How much is this in Japanese yen?

....

Skills Survey

Add the scores in Exs. 1-4. Then divide each total by the number of scores added to get the average. Round each answer to the nearest whole number.

1.	**2.**	**3.**	**4.**
5	34	125	95
6	45	180	110
4	36	155	124
3	42	200	87
2	27	+ 160	106
6	+ 113		100
+ 4			+ 98

Arrange the numbers in Exs. 5-7 from greatest to least in value.

5.		**6.**		**7.**	
1.00	....	0.600	....	1.009	....
0.02	....	0.537	....	0.958	....
3.20	....	0.421	....	1.010	....
0.12	....	0.708	....	0.957	....
4.09	....	0.375	....	1.101	....
4.25	....	0.675	....	0.897	....
3.40	....	0.500	....	1.001	....
1.60	....	0.357	....	1.210	....
0.50	....	0.676	....	0.960	....

Circle the operation or operations you need to use for each problem in Exs. 8-11.

8. If a batsman hits four sixes in a match, how many runs does he make?
ADD SUBTRACT MULTIPLY DIVIDE

9. What is the total measurement of a bag that is 50 cm high, 65 cm wide and 15 cm deep?
ADD SUBTRACT MULTIPLY DIVIDE

10. What is your batting average if you score 120, 130 and 110 in three games?
ADD SUBTRACT MULTIPLY DIVIDE

11. In twenty-nine games the Blues won twenty-nine. How many games did they lose?
ADD SUBTRACT MULTIPLY DIVIDE

Use the following formulae to solve problems 12 and 13:

$$\text{Distance} = \text{Speed} \times \text{Time} \qquad \text{Time} = \frac{\text{Distance}}{\text{Speed}} \qquad \text{Speed} = \frac{\text{Distance}}{\text{Time}}$$

12. In 5 hrs you were able to drive 250 m. How fast were you going?

13. You drove 100 m at 50 m.p.h. How long did it take you?

Branching Off

★ The time of day varies in different parts of the world. If it's 7.00 a.m. in London, what time is it in Singapore, Paris, Sydney, Madrid, Israel, Hawaii and San Francisco?
Find a time zone map in a diary or an almanac.

★ Find out how to score a ten-pin bowling game.

UNIT 7

This section contains exercises and facts to help you apply your maths skills to real-life problems.

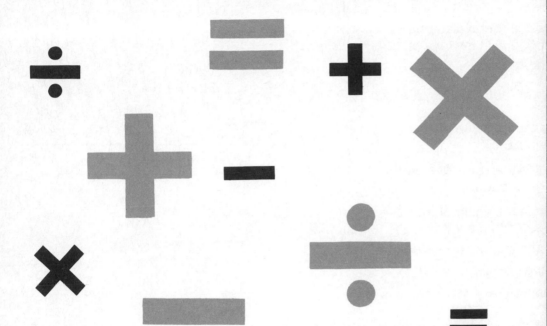

Basic Skills and Reference

Place Value

Whole Numbers

The value of **5** in each of the places shown on the chart is different. Each place has *ten times the value* of the next place to the right. **5** in the hundreds place has a value of 5 × 100, or 500. **5** in the tens place has a value of 5 × 10, or 50.

a. Answer the following questions:

1. What is the value of six in the tens place? ....

2. What is the value of seven in the ones place? ....

3. What is the value of four in the thousands place? ....

4. If eight is in the ten thousands place, its value is: ....

5. 3000 means that the three is in the .... place.

b. Write these numbers in their expanded form:

842 = 800 + 40 + 2   1030 = 1000 + 30

9409 = ... 3015 = ...

10450 = ... 7912 = ...

25527 = ...

c. Write the following numerals in this place value chart:

PLACE VALUE

Hundred Millions	Ten Millions	Millions	Hundred Thousands	Ten Thousands	Thousands	Hundreds	Tens	Ones
5	5	5	5	5	5	5	5	5

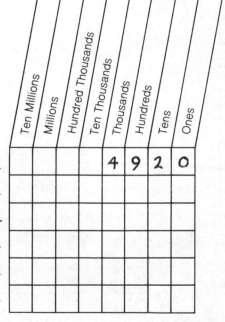

PLACE VALUE

Four thousand, nine hundred and twenty ➞
Two hundred and thirty four ➞
Fifteen thousand, seven hundred and ten ➞
Nine thousand and three ➞
One million, two hundred and seven thousand ➞
Ninety three thousand and sixty five ➞

Ten Millions	Millions	Hundred Thousands	Ten Thousands	Thousands	Hundreds	Tens	Ones
				4	9	2	0

Decimal Numbers

On this chart, the value of each 5 after the decimal point is different. Each place has ten times the value of the next place to the right. **5** in the tenths place has a value of $^5/_{10}$ or 0.5. **5** in the hundredths place has a value of $^5/_{100}$ or 0.05.

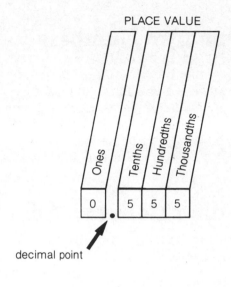

d. Answer the following questions:

1. What is the value of four in the tenths place? ....

2. $\frac{3}{100}$ is another name for three in the

.... place.

3. 0.07 means that seven is in the

.... place.

4. What is the value of two in the thousandths place? ....

5. 0.0009 means that nine is in the

.... place.

e. Write the following numerals in this place value chart.

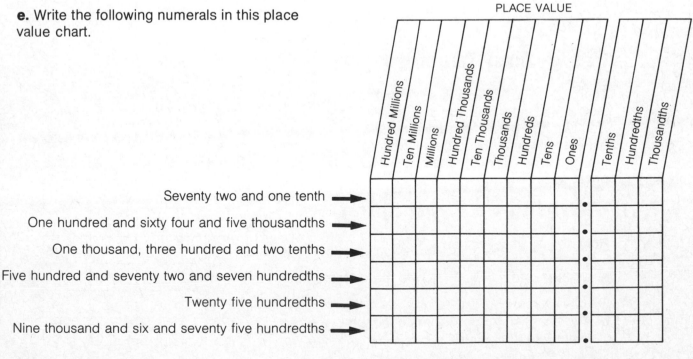

Addition

Lining Up Numbers to Add

Line up numbers by place value. Ones must line up with ones, tens must line up with tens, and so on. To add 23 + 1 + 3251 + 401, line them up this way:

```
  23
   1
3251
 401
```

Line up these numbers: **1.** 235 + 4 + 61 + 4000

2. 4312 + 34 + 5 + 789

Adding Columns from Right to Left

Add the ones.

```
301
452
  3
```

3.
a.
```
125
243
```

4.
a.
```
617
321
```

5.
a.
```
223
132
```

Add the tens.

```
301
452
 5
```

b.
```
124
243
```

b.
```
617
321
```

b.
```
223
132
```

Add the hundreds.

```
301
452
7
```

c.
```
125
243
```

c.
```
617
321
```

c.
```
223
132
```

Write the sum.

753   d ....   d ....   d ....

Hundreds	Tens	Ones
3	4	5
300	40	5

Using Your Memory

To find 5 + 3 + 9, you first think of 5 + 3 = 8, and then add 9 to get 17. Add two numbers first, remember the sum, and then add another number to the remembered sum, and so on. The order does not matter.

Find the sums.

6. 2 + 1 + 5 = ...

7. 3 + 4 + 6 = ...

8. 4 + 5 + 2 + 1 = ...

9. 1 + 3 + 7 + 9 = ...

10. 6 + 5 + 8 + 7 = ...

Carrying or Regrouping

How do you usually add 68 + 26?

a. Add the ones, 8 + 6 =; write the 4. Remember 1 ten from 14

```
  68
+ 26
   4
```

b. Add the remembered 1 ten to the tens, 1 + 6 + 2 = 9. Write 9 to the left of 4.

```
  68
+ 26
  94
```

110

Find the sums.

11.	¹74 + 16 90	57 + 38	26 + 55	49 + 48	**14.**	¹126 + 59 185	782 + 156	365 + 809	485 + 760
12.	¹78 + 36 114	85 + 75	49 + 68	39 + 84	**15.**	¹¹924 + 76 1000	864 + 247	346 + 876	984 + 249
13.	¹¹175 + 28 203	29 + 384	354 + 296	158 + 493	**16.**	¹1787 + 907 2694	2528 + 645	2637 + 7363	888 + 888

Adding Long Columns

You can add long columns of numbers in different ways. Study the following methods:

Method 1
Add each column and carry the tens to the next column.

```
 ¹ ¹
  455
  658
  834
  212
 2159
```

Method 2
```
  455
  658
  834
+ 212
```
```
   19  ← add the ones
  140  ← add the tens
 2000  ← add the hundreds
 2159
```

Method 3
```
  455
  658   1113 (partial sum)
  834
+ 212   1046 (partial sum)
        2159 (sum)
```

Method 4
First collect numbers that add up to ten, then add the other numbers in the column.

Find the totals using the method that's easiest for you.

17.	124 953 687 + 456	**18.**	875 235 492 + 618	**19.**	267 725 128 + 953	**20.**	786 935 547 109 + 63

Checking Sums

One way to check your answer in addition is to change the order of the numbers. You should get the same sum. Another way is to add first from the bottom and then check by adding down from the top. Check your answers in Exs. 17-20.

Adding Decimals

This is done in the same way as whole numbers, but you must line up the decimal points

```
20.43 + 105.3      20.43
                 + 105.3
                  125.73
```

```
145.02 + 16.98    145.02
                + 16.98
                 162.00
```

Line up the decimal points and find the totals.

21. 10.75 + 24.3

22. 126.25 + 27.9 + 6

23. 7.96 + 3.04

24. 74 + 12.95 + 3.06

Subtraction

Addition and Subtraction

Addition

```
    2
  + 3
```

```
    ?
  + 3
```

Subtraction

```
    5
  - 3
```

Rewrite each item into a subtraction problem. Then find the difference.

1.
```
    ?
  + 7
  ----
   18
```

2.
```
   22
  + ?
  ----
   28
```

3.
```
    ?
  + 7
  ----
   39
```

4.
```
   35
  + ?
  ----
   47
```

5.
```
    ?
 + 213
 -----
  568
```

6.
```
  701
  + ?
 -----
  829
```

Renaming Numbers in Subtraction

How do you work out 45 − 18?

```
   45
 - 18
 ----
    ?
```

a. You can't take 8 from 5, so rename the 4 tens in 45 as 3 tens + 1 ten.

b. Add the 1 ten to the 5 in the ones column (top line): 10 + 5 = 15.

c. Then 15 − 8 = 7. Write 7.   D. In the tens column, 3 − 1 = 2. Write 2.

```
   3 15
   4 5
 -  18
 -----
   27
```

Subtraction Using Equal Addition

You may use a different method of subtraction to work out 45 − 18.

a. You can't take 8 from 5; add 1 ten to 5 in the ones place (top line); 10 + 5 = 15

b. Then 15 − 8 = 7. Write 7.

c. Add 1 ten to 1 in the tens place (lower line)

d. In the tens column, 4 − 2 = 2. Write 2.

```
     15
    4 5
 - 2 18
 ------
    27
```

(When you add equally to the top and to the lower lines, the answer stays the same.)

Use one of these methods to find the differences.

7.   22   8.   35   9.   26   10.   47   11.   58
    − 3   − 8   − 7   − 9   − 9

12. 431   13. 352   14. 234   15. 655   16. 286
    − 63   − 74   − 55   − 66   − 98

17. 222   18. 425   19. 512   20. 356   21. 685
   − 169   − 358   − 424   − 267   − 598

22. 300   23. 5200   24. 2005   25. 1050   26. 3020
    − 23   − 199   − 576   − 561   − 342

Checking the Difference

One way of checking your answer to a subtraction problem
is to add the difference and the lower number. The sum
should be equal to the larger number. Check your
answers in this section, using this method.

```
 1260          1073
−  187    ⤫  +  187
 1073          1260
```

Subtracting Decimals

This is done in the same way as whole numbers, but you must line up the decimal points.

$$9.7 - 6.4$$
```
   9.7
 − 6.4
   3.3
```

$$27.25 - 15.9$$
```
  6 12                    11
 27.25                  27.25
− 15.9      or        − 15.9
 11.35                  11.35
```

Line up the decimal points and find the differences.

27. 124.5 − 96.3   28. 7.9 − 1.24

29. 28.59 − 17.08   30. 12 − 7.63

Multiplication

Multiplying from Right to Left

How do you usually work out 423×2?

a. $3 \times 2 = 6$; write 6.
b. $2 \times 2 = 4$; write 4 to the left of 6.

c. $4 \times 2 = 8$; write 8 to the left of 4.

$$\begin{array}{r} 423 \\ \times\ 2 \\ \hline 846 \\ \textbf{cba} \end{array}$$

Find the products.

1.	2.	3.	4.
$\begin{array}{r} 23 \\ \times\ 3 \\ \hline \end{array}$	$\begin{array}{r} 385 \\ \times\ 1 \\ \hline \end{array}$	$\begin{array}{r} 301 \\ \times\ 2 \\ \hline \end{array}$	$\begin{array}{r} 72 \\ \times\ 4 \\ \hline \end{array}$

5.	6.	7.	8.
$\begin{array}{r} 80 \\ \times\ 7 \\ \hline \end{array}$	$\begin{array}{r} 511 \\ \times\ 6 \\ \hline \end{array}$	$\begin{array}{r} 802 \\ \times\ 4 \\ \hline \end{array}$	$\begin{array}{r} 931 \\ \times\ 3 \\ \hline \end{array}$

Using your Memory in Multiplication

How do you usually work out 87×4?
$7 \times 4 = 28$, write 8. $\longrightarrow$

$$\begin{array}{r} 87 \\ \times\ 4 \\ \hline 8 \end{array}$$

Remember 2 from 28.
  $8 \times 4 = 32$, add the remembered 2,
$32 + 2 = 34$
Write 34 to the left of 8. $\longrightarrow$

$$\begin{array}{r} 87 \\ \times\ 4 \\ \hline 348 \end{array}$$

Find the products.

9.	10.	11.
$\begin{array}{r} 95 \\ \times\ 6 \\ \hline \end{array}$	$\begin{array}{r} 87 \\ \times\ 5 \\ \hline \end{array}$	$\begin{array}{r} 64 \\ \times\ 8 \\ \hline \end{array}$

12.	13.	14.
$\begin{array}{r} 137 \\ \times\ 2 \\ \hline \end{array}$	$\begin{array}{r} 209 \\ \times\ 4 \\ \hline \end{array}$	$\begin{array}{r} 514 \\ \times\ 7 \\ \hline \end{array}$

Multiplying by 10, 100, 1000

To multiply a number by 10, 100, or 1000,
here's what you do:

$$\begin{array}{rl} 35 \times 10 = & 350 \\ 35 \times 100 = & 3500 \\ 35 \times 1000 = & 35000 \end{array}$$

Find the products.

15. $58 \times 10 =$ ....

16. $58 \times 100 =$ ....

17. $58 \times 1000 =$ ....

18. $60 \times 100 =$ ....

19. $45 \times 10 =$ ....

20. $99 \times 1000 =$ ....

21. $125 \times 100 =$ ....

Using Two Partial Products

21. To find 27×56, you often use the
following method.

$$\begin{array}{rl} 27 & \\ \times\ 56 & \\ \hline 162 & \leftarrow (27 \times 6) \text{ partial product} \\ +\ 1350 & \leftarrow (27 \times 50) \text{ partial product} \\ \hline 1512 & \leftarrow (162 + 1350) \text{ PRODUCT} \end{array}$$

Find the products.

22.	23.	24.
$\begin{array}{r} 42 \\ \times\ 23 \\ \hline \end{array}$	$\begin{array}{r} 46 \\ \times\ 31 \\ \hline \end{array}$	$\begin{array}{r} 81 \\ \times\ 19 \\ \hline \end{array}$

25.	26.	27.
$\begin{array}{r} 132 \\ \times\ 24 \\ \hline \end{array}$	$\begin{array}{r} 345 \\ \times\ 63 \\ \hline \end{array}$	$\begin{array}{r} 1213 \\ \times\ 32 \\ \hline \end{array}$

Using Three Partial Products

To find the product of 692 × 231, you often use the following method:

```
      692
   ×  231
      692   ← (692 × 1) partial product
    20 760   ← (692 × 30) partial product
 + 138 400   ← (692 × 200) partial product
  159852   ← PRODUCT
```

Find the products.

28. 765   **29.** 348
      × 211   × 123

30. 879   **31.** 647
      × 312   × 251

Zeros in Multiplication

Your solution to 225 × 304 may be written in two different ways:

First Method: Second Method:

```
      225                 225
   ×  304              ×  304
      900                 900
    0 000              67 500
  + 67 500             68 400
   68 400
```

Use either method to find the products.

32. 352   **33.** 864
      × 205   × 302

34. 506   **35.** 708
      × 201   × 403

Checking Your Answers

One way of proving your product is to interchange the two numbers to be multiplied.

```
                     Check:
      43                 15
   ×  15              ×  43
     215                 45
     430                600
     645  ← PRODUCT →    645
```
Prove each product in Exs. 32–35.

Multiplying Decimals by Whole Numbers

This is done in the same way as multiplying whole numbers by whole numbers, but you must put the decimal point in the right place.

```
0.07 × 6     0.07        2.75 × 13     2.75
            ×   6                    ×   13
            0.42                      8.25
                                    27.50
                                    35.75
```

Find the products.

36. 2.91 × 5 = ...

37. 12.06 × 10 = ...

38. 146.99 × 2 = ...

39. 3.75 × 15 = ...

Division

Solving Division Problems

$276 \div 23$ is usually solved this way:

$$
\begin{array}{r}
12 \\
23 \overline{)276} \\
\end{array}
$$

$(10 \times 23) \longrightarrow$ 230

$(2 \times 23) \longrightarrow$ 46

$$
\begin{array}{r}
46 \\
\hline
00 \\
\end{array}
$$

Find the quotients:

1. $9\overline{)828}$

2. $18\overline{)234}$

Zero in the Quotient

The answer to $2461 \div 23$ is sometimes incorrectly written as 17. It should be 107. To avoid this error, you may write your work like this:

$$
\begin{array}{r}
107 \\
23 \overline{)2461} \\
23 \\
\hline
16 \\
00 \\
\hline
161 \\
161 \\
\hline
000 \\
\end{array}
$$

Remember: Each time you bring down one digit from the dividend, you must write one digit in the quotient.

You may also avoid the mistake by estimating or guessing the quotient.

Estimate $2461 \div 23$.

Round 2461 to 2000 and 23 to 20.

Since $2000 \div 20 = 100$, you know that the quotient is about 100. So 17 is wrong.

Find the quotients:

3. $32\overline{)6592}$

4. $19\overline{)5795}$

5. $24\overline{)9696}$

6. $17\overline{)8534}$

Short Method of Dividing Rounded Numbers

In multiplying 200 by 20, you simply write three zeros and multiply 2 by 2. Your answer is 4000.

To divide 4000 by 200, this is what you do:

$40\cancel{00} \div 2\cancel{00} = 40 \div 2 = 20$

To divide 8000 by 4000:

$8\cancel{000} \div 4\cancel{000} = 8 \div 4 = 2$

Find the quotients:

7. $6000 \div 2000 =$ ....

8. $4500 \div 900 =$ ....

9. $3500 \div 700 =$ ....

10. $800 \div 200 =$ ....

11. Write a simple rule for dividing rounded numbers ....
....

What will be the first digit in each quotient?

12. 40)1728   48)1728

13. 70)3672   72)3672

14. 20)1152   24)1152

Check if each answer is reasonable.

15. $\begin{array}{r} 47 \\ 35)\overline{1645} \end{array}$ ....

16. $\begin{array}{r} 57 \\ 71)\overline{4047} \end{array}$ ....

17. $\begin{array}{r} 34 \\ 27)\overline{8208} \end{array}$ ....

Find each quotient.

18. 72)3672   **19.** 24)1152

Remainders in Division

To change minutes to hours, you divide the number of minutes by 60 (60 min = 1 hr). Sometimes there are leftover minutes. In division, these leftovers are called remainders.

135 min ÷ 60 = ?
The answer is 2 hours and 15 minutes

$$\begin{array}{r} 2 \\ 60)\overline{135} \\ 120 \\ \overline{15} \end{array} \leftarrow \text{Remainder}$$

*Do **not** use your calculators for these exercises. The calculator gives you fractions of an hour, not minutes.*

Find the quotients and remainders:

20. 198 min ÷ 60 = ... hours .... minutes

21. 59 days ÷ 7 = ... weeks .... days

22. 127 months ÷ 12 = ... years .... months

23. 173 cm ÷ 100 = ... metres .... centimetres

Dividing Decimals by Whole Numbers

This is done in the same way as dividing whole numbers by whole numbers, but you must put the decimal point in the right place.

$$0.75 \div 5 \quad \begin{array}{r} 0.15 \\ 5)\overline{0.75} \\ 0 \\ \overline{7} \\ 5 \\ \overline{25} \\ 25 \\ \overline{00} \end{array} \qquad 86.4 \div 24 \quad \begin{array}{r} 3.6 \\ 24)\overline{86.4} \\ 72 \\ \overline{14\,4} \\ 14\,4 \\ \overline{00\,0} \end{array}$$

Find the quotients:

24. 57.6 ÷ 12 = ...

25. 11.83 ÷ 7 = ...

26. 107.25 ÷ 3 = ...

27. 46.5 ÷ 25 = ...

Addition/Subtraction Table

	0	1	2	3	4	5	6	7	8	9	10
0	0	1	2	3	4	5	6	7	8	9	10
1	1	2	3	4	5	6	7	8	9	10	11
2	2	3	4	5	6	7	8	9	10	11	12
3	3	4	5	6	7	8	9	10	11	12	13
4	4	5	6	7	8	9	10	11	12	13	14
5	5	6	7	8	9	10	11	12	13	14	15
6	6	7	8	9	10	11	12	13	14	15	16
7	7	8	9	10	11	12	13	14	15	16	17
8	8	9	10	11	12	13	14	15	16	17	18
9	9	10	11	12	13	14	15	16	17	18	19
10	10	11	12	13	14	15	16	17	18	19	20

$$\begin{array}{r} 5 \\ +7 \\ \hline 12 \end{array} \qquad \begin{array}{r} 17 \\ -9 \\ \hline 8 \end{array}$$

The answers to the following problems can be found in the table. Write the letter of the problem next to the answer.

a. $13 - 6 = ?$ **b.** $8 + 7 = ?$ **c.** $14 - 9 = ?$

Multiplication/Division Table

	1	2	3	4	5	6	7	8	9	10
0	0	0	0	0	0	0	0	0	0	0
1	1	2	3	4	5	6	7	8	9	10
2	2	4	6	8	10	12	14	16	18	20
3	3	6	9	12	15	18	21	24	27	30
4	4	8	12	16	20	24	28	32	36	40
5	5	10	15	20	25	30	35	40	45	50
6	6	12	18	24	30	36	42	48	54	60
7	7	14	21	28	35	42	49	56	63	70
8	8	16	24	32	40	48	56	64	72	80
9	9	18	27	36	45	54	63	72	81	90
10	10	20	30	40	50	60	70	80	90	100

$$\begin{array}{r} 8 \\ \times 7 \\ \hline 56 \end{array} \qquad \begin{array}{r} 6 \\ 9\overline{)54} \end{array}$$

The answers to the following problems can be found in the table. Put the letter of the problem next to the answer.

d. $36 \div 9 = ?$ **e.** $7 \times 9 = ?$ **f.** $42 \div 6 = ?$

Estimating

Guessing the Answer

"Have I got enough money?" How often do you ask yourself this just before going to the cashier? You can find a quick, reasonable answer by estimating – guessing a fairly close answer to a maths problem. Estimating can be done in many ways. It will give you a rough check of your answers.

Rounding

A first step in estimating is to round numbers:

76 rounded to the nearest ten ......80......

71 rounded to nearest ten ......70......

127 rounded to the nearest ten 130......

373 rounded to the nearest hundred ......400......

7543 rounded to nearest hundred ......7500......

Round these numbers to the nearest ten:

1. 399 ....   2. 207 ....

3. 964 ....   4. 72 ....

Round these numbers to the nearest hundred:

5. 156 ....   6. 973 ....

7. 2591 ....   8. 7035 ....

Round these numbers to the nearest thousand:

9. 896 ....   10. 3550 ....

11. 9015 ....   12. 6999 ....

Checking Answers by Estimating

By guessing the answer to a problem, you have a way of checking if your actual answer is right or wrong. For example, if your estimate is 1000 and your actual answer is 110, you know that you have made a mistake somewhere. You can then do the problem again.

In maths problems, if you round numbers to the nearest ten, hundred, or thousand, you can work with them mentally. To estimate 898 + 204, for example:

$$\begin{array}{llr} 898 & \text{is rounded to} & 900 \\ +\ 204 & \text{is rounded to} & +\ 200 \\ \hline Total \rightarrow 1102 & & 1100 \leftarrow Estimate \end{array}$$

Estimate the totals:

13. 813 + 692: .... + ... = ...

14. 3185 + 1812: .... + ... = ...

15. 62 + 78 + 39: .... + ... + ... = ...

Estimate the differences:

16. 706 − 598: .... − .... = ...

17. 497 − 208: .... − .... = ...

18. 6028 − 3982: .... − .... = ...

Estimate the products:

19. 29 × 31: .... × .... = ...

20. 88 × 52: .... × .... = ...

21. 394 × 203: .... × .... = ...

Estimate the quotients:

22. 4105 ÷ 79: .... ÷ .... = ...

23. 2950 ÷ 51: .... ÷ .... = ...

Using Estimates Every Day

When you need to pay for things, estimating can help you work out how much money you will need.

The question to be answered often tells you how to estimate. For example, here are two different questions about the same ad. The estimates are done in different ways.

What is the cost of two batteries?

Step 1: Round 38p to 40p

Step 2: Calculate 40p + 40p or 40p × 2 to get the answer – about 80p.

Is 80p enough to buy two batteries?

Step 1: Calculate 80p ÷ 2 = 40p

Step 2: Think, since 38p is about 40p the answer is most likely to be "yes".

Look at the facts in each ad. Quickly guess the answer to each question and circle it. Don't do any calculations on paper or by calculator.

BUTTONS

29p each

24. How much will two buttons cost?

   About 60p   About 40p

25. Is 90p enough to buy three buttons?

   Yes   No

26. If I buy one button, how much change will I get from 50p

   About 30p   About 20p

CASSETTES
3 for £5

27. How much does each cassette cost?

   About £1.00   About £2.00

28. Can you buy two cassettes for £4.00?

   Yes   No

29. How many cassettes can £12.00 buy?

   6   8   10

SHIRTS £8·99
TROUSERS £11·99

30. You have £20.00. Is that enough to buy a shirt and a pair of trousers?

   Yes   No

31. If you buy two shirts, how much will they cost?

   About £16.00   About £18.00

32. Can you buy two pairs of trousers for £20.00?

   Yes   No

Finding Percentages

A percentage is a kind of fraction. Percentages are often used in business to find interest, discount, mark-up and commission.

★ "Percent" means rate per 100.

Five percent means 5 of every 100 or $\frac{5}{100}$

This is written as 5%.

$$\frac{5}{100} = 0.05$$

$$
\begin{array}{r}
0.05 \\
100\overline{\smash{)}5.00} \\
0 \\
\hline
5\,0 \\
0 \\
\hline
5\,00 \\
5\,00 \\
\hline
0\,00 \\
\end{array}
$$

★ To find a percentage of a number:
    3% of 64

Write the percent as a decimal. **3.%**

Drop the percent sign. **3.**

Move the decimal point 2 places to the left of its original position. **0.03**

Multiply the number by the decimal.

$$
\begin{array}{lr}
64.00 & \textit{(2 decimal places)} \\
\times\ \ 0.03 & \textit{(2 decimal places)} \\
\hline
1.9200 & \textit{(4 decimal places)} \\
\end{array}
$$

3% of 64 = 1.92

★ To find a percentage of an amount, follow the steps above.

    5% of £4.00
    5% = 0.05

$$
\begin{array}{lr}
£4.00 & \textit{(2 decimal places)} \\
\times\ \ 0.05 & \textit{(2 decimal places)} \\
\hline
.2000 & \textit{(4 decimal places)} \\
\end{array}
$$

= £0.20

Work out the following percentages:

1. 1% of 15 = 0.15

$$
\begin{array}{lr}
15.00 & \textit{(2 decimal places)} \\
\times\ \ 0.01 & \textit{(2 decimal places)} \\
\hline
0.1500 & \textit{(4 decimal places)} \\
\end{array}
$$

2. 10% of 84   = ...

3. 40% of 120   = ...

4. 25% of 600   = ...

5. 15% of 1000   = ...

6. 12½% of 48   = ...

7. 3% of £25.00   = ...

8. 4% of £666.00   = ...

9. 5% of £7.00   = ...

10. 6% of £16.00   = ...

11. 20% of £350.00   = ...

12. 15% of £7000.00   = ...

★ Sometimes you need to work out what percentage one amount is of another. What percentage is £2.00 of £16.00?

First make a fraction: $\dfrac{£2.00}{£16.00}$

Drop the pound sign: $\dfrac{2}{16}$

Find the simplest name for the fraction: $\dfrac{2}{16} = \dfrac{1}{8}$

Change the fraction to a decimal: $\dfrac{1}{8} = 0.125$

Multiply this decimal by 100 to give percent.

    0.125 × 100 = 12.5 or 12½%

13. What percentage is £2.00 of £8.00?   ....

14. What percentage is £0.05 of £10.00?   ....

15. What percentage is £2.50 of £25.00?   ....

16. What percentage is £125.00 of £1000.00?   ....

You can use your calculator for percentages.
(See p. 12)

Fractions, Decimals and Percentages Table

Fraction		Decimal		Percentage
$1\left(\frac{1}{1}\right)$	=	1.0	=	100%
$2\left(\frac{2}{1}\right)$	=	2.0	=	200%
$2\frac{1}{2}$	=	2.5	=	250%
$\frac{1}{100}$	=	0.01	=	1%
$\frac{1}{10}$	=	0.1	=	10%
$\frac{1}{20}$	=	0.05	=	5%
$\frac{1}{2}$	=	0.5	=	50%
$\frac{1}{4}$	=	0.25	=	25%
$\frac{3}{4}$	=	0.75	=	75%
$\frac{1}{5}$	=	0.2	=	20%
$\frac{2}{5}$	=	0.4	=	40%
$\frac{3}{5}$	=	0.6	=	60%
$\frac{4}{5}$	=	0.8	=	80%
$\frac{1}{8}$	=	0.125	=	$12\frac{1}{2}$%
$\frac{3}{8}$	=	0.375	=	$37\frac{1}{2}$%
$\frac{5}{8}$	=	0.625	=	$62\frac{1}{2}$%
$\frac{7}{8}$	=	0.875	=	$87\frac{1}{2}$%
$\frac{1}{3}$	=	0.333...	=	$33\frac{1}{3}$%
$\frac{2}{3}$	=	0.666...	=	$66\frac{2}{3}$%

Using Metric Measures

Metre, litre and **gram**—these are the basic units for measuring length, capacity (volume) and mass (weight). Metric measurement is based on ten. It is a decimal system using many standard prefixes as shown on the chart below.

The Common Prefixes and Units

Prefix	kilo	(unit)	deci*	centi*	milli
Symbol	k	(m, L or g)	d	c	m
Decimal value	1000	1	0.1	0.01	0.001
Meaning	a thousand	one	one tenth	one hundredth	one thousandth

*centi and **deci** are used only with the metre

Use the information on the prefix chart to complete this table.

Name of Unit	Symbol	Change to	Operation	Example
millimetre		cm	÷ 10	40mm = ... cm
	cm	mm	× 10	2 cm = ... mm
metre	m		× 100	3 m = ... cm
metre	m	km	÷ 1000	5000 m = ... km
kilometre		m		60 km = ... m
kilogram		g	× 1000	5 kg = ... g
gram			÷ 1000	2000 g = ... kg
	mg	g	÷ 1000	4000 mg = ... g
	g	mg		3 g = ... mg
litre		kl	÷ 1000	1200 l = ... kl
	ml	l	÷ 1000	4500 = ... l

How long?

millimetre (mm)
centimetre (cm)
metre (m)
kilometre (km)
10 mm = 1 cm
100 cm = 1 m
1000 m = 1 km

1 mm	About the thickness of a ½p coin
2 mm	About the thickness of a match stick
1 cm	About the width of an index finger nail
10 cm	About the width of a man's fist
1 m	A long pace
50 m	Length of an Olympic swimming pool

How Big?

square millimetre (mm²)
square centimetre (cm²)
square metre (m²)
hectare (ha)
square kilometre (km²)

100 mm²	=	1 cm²
10000 cm²	=	1 m²
10000 m²	=	1 ha
100 ha	=	1 km²

2 mm²   About the area of the top of a pin head

5 cm²   About the area of a postage stamp

1 m²   About the area of a shower recess floor

100 m²   About the area of a small family house

1000m²   The area of an Olympic swimming pool (50 m × 20 m)

1ha   About the area of a football field.

How Full?

millilitre (ml)
Litre (l)
cubic centimetre (cm³)
cubic metre (m³)
1 cubic metre (m³) = 1000 litres (l)
1 litre (l) = 1000 millilitres (ml)
1 cubic centimetre (cm³) = 1 millilitre (ml)

1 ml   About one eye-dropper full

5 ml   One standard teaspoon

200 ml   About a glass full

1 l   Large bottle of cola

8 l   About average bucket

How Heavy?

gram (gram)
kilogram (kg)
tonne (t)
1000 g = 1 kg
1000 kg = 1 tonne

1 g   About the mass (weight) of three aspirin tablets

5 g   About the mass of a 5p piece

50 g   About the mass of a golf ball

500 g   About the mass of eight thick sausages

1 kg   About the mass of eight medium size apples

20 kg   Luggage allowance for economy class air travel

70kg   About the mass of an average man

1 t   About the mass of a small lorry

How Fast?

kilometre per hour (km/h)

5 km/h   Average speed for walking

6 km/h   Average speed for swimming 100 m in one minute

50 km/h   Speed limit in built-up areas

850 km/h   Normal cruising speed of jet plane

How Hot? How Cold?

degree Celsius (°C)

100°C	Boiling point of water
37°C	Body temperature
30°C	Very hot weather
25°C	Hot weather
20°C	Warm weather
15°C	Mild weather
10°C 5°C	Cold to very cold weather
0°C	Freezing point of water

Oven Temperatures

	Very Cool	Cool	Moderate	Hot	Very Hot
°C	110-114	150-160	180-190	220-230	250-260

Body Temperatures

	Normal	Feverish	Fever	High Fever
°C	37	38	39	40

Hints about writing metric measures.

★ Do not use full stops after metric symbols: 7 kg

★ Leave a small space between the number and the symbol: 25 l

★ Do not add 's' to symbols: 1 g 10 g

★ Do not use commas to separate groups of numerals. Use a small space to separate groups of three numerals **in numbers containing more than four numerals:**

2000   7350   10 500   7 396 250

Answer the following questions:

Which unit is often used to measure fabric?   ....

Which unit is used to measure distances between cities?   ....

Petrol is measured in   ....

Airmail letters would be weighed in   ....

To find out how tall you are, which unit would you use?   ....

A dose of liquid medicine might be shown in   ....

Oven heat is shown in   ....

The size of a tile is often shown in   ....

Which unit is used on tins of fruit juice?   ....

The measurement of an area of land might be shown in square   ....

Glossary

account—The record of one's money in a bank.

accuracy—The exactness or correctness of an answer.

adaptor—A transformer that allows a battery-operated appliance to use household electricity.

addition—The process of adding numbers to get a sum.

amount—The quantity or number of something.

area—The number of unit squares on a surface (length multiplied by width).

average—A number equal to the sum divided by the number of units which make up the sum.

balance—The amount of money remaining in an account after a deposit has been added or a payment has been subtracted.

bank statement—A personal record of a cheque account showing cheques written and deposits made.

bookkeeping—The method of recording the income and expenses of a business.

budget—A plan that helps make the best use of money or time.

calculate—To find an answer by adding, subtracting, multiplying or dividing.

calculator—An electronic instrument used to work out maths problems.

cash—Money that is immediately available to spend.

cash record—A statement that shows the balance after adding amounts received or subtracting amounts paid out.

Celsius—A scale of temperature on which water freezes at 0° (zero degrees) and boils at 100° (100 degrees).

chart—A drawing, diagram, map or table that puts together information so that each item can be easily found.

cheque—A written order for the bank to pay money from an account to the person named on the cheque.

commission—A method of payment based on a percentage of a saleperson's total sales.

commuter—A person who travels regularly to and from work.

compute—To calculate the answer to a maths problem.

consumer—A buyer of goods or services.

consumption—The amount used.

credit—A loan or borrowed amount to be paid back after the promised period of time.

current account—A bank account from which you withdraw money by writing cheques.

decimal—A special kind of fraction based on tenths.

decimal points—A dot placed before a fraction given in decimal figures, as in 2.03 or 0.623.

deduct—To take away or subtract.

deduction—The amount which is taken away (usually from gross pay or from taxable income).

denominator—The number below the line in a fraction.

deposit—To put money in a bank account.

difference—The answer to a subtraction problem.

digit—Any of the figures 0, 1, 2, 3, 4, 5, 6, 7, 8, and 9 which make up numerals.

discount—The amount taken off the usual price.

dividend—The number which is to be divided by another number.

division—The process of separating something into a number of parts.

equivalent—Equal to.

estimate—To make a reasonable guess.

excess—More than that allowed.

expense—An amount paid out.

finance charge—Interest or amount paid in addition to the amount borrowed.

finance company—A company which lends money and charges interest on those loans.

fixed expenses—Amounts to pay which are the same, or nearly the same, each month.

flexible expenses—Amounts to pay which may vary more than fixed expenses, or do not occur each month.

fraction—A part of a whole expressed as a number with a numerator and a denominator.

fractional cost—The cost of part of the whole.

graph—A representation in picture form of related facts or figures.

gross earnings, income or pay—The total amount earned before any deductions are subtracted.

gross profit—The total amount earned by a business before expenses are deducted.

income—Money earned from work or from business.

income tax—The tax paid on a person's (or business's) net income.

insurance—An agreement that arranges for money payment in the case of loss, accident, fire or death.

interest—A percentage paid on an amount of money borrowed or a percentage earned on an amount of money deposited.

loan—Money lent to a borrower for an agreed time. Interest is usually charged.

loss—The difference between income and expenses when the expenses are greater than the income.

mark-up—An amount added to the unit cost to find the selling price.

maximum—The largest or highest amount.

memory—The ability to store and recall information.

metric—Referring to the metric system (a decimal measuring system which includes metre, litre and kilogram as basic units).

minimum—The least or lowest amount.

mortgage—The loan made against the security of a home.

multiplication—The process of multiplying one number by another.

multiply—To add a number to itself a given number of times.

net earnings, income or pay—The amount of earnings left after deductions have been made; 'take-home' pay.

net loss—The amount of money lost when operating expenses exceed profit.

net profit—The amount of money gained after all operating expenses have been paid.

new balance—The balance which appears in a record after an amount has been added or subtracted.

nil—No amount.

number—The sum or count of a group of things or people.

numeral—A figure or a group of figures standing for a number.

numerator—The number above the line in a fraction.

operating expenses—The amount of money needed to run a business (this includes expenses for rent, wages, electricity, advertisements, supplies, transport, etc).

original—The first.

overtime—Time worked in excess or normal hours.

p.a.—Per annum (each year).

partial—Forming only a part; not complete.

passbook—A record of deposits and withdrawals in a savings account.

payments—Amounts that are paid.

per—For each.

percent, percentage—Part of each hundred.

piece rate—The amount of money earned on each piece made.

piecework earnings—Income calculated by multiplying the piece rate by the number of pieces made.

place value—The value based on the position of a digit in a numeral.

product—The answer to a multiplication problem.

profit—The amount of money that is left after operating expenses are subtracted from gross earnings.

purchases—Things which have been paid for.

quantity (qty)—The number of items bought or sold.

quotient—The answer to a division problem.

rate—A set value.

record of purchases—A list of items bought by a business.

remainder—The number which is left after subtraction or division.

rounded numbers—Numbers which have been changed, usually to the nearest whole number, ten or hundred.

sale—The selling of goods at discounted (reduced) prices.

sales report—A record of the total income of a business over a given period of time.

savings account—A bank account in which money is deposited for safekeeping and for earning interest.

stock—The value of all the goods for sale.

S.T.D.—Subscriber Trunk Dialling; trunk calls in which the caller dials the number directly without help from an operator.

subtraction—The process of finding the difference between two numbers.

sum—The result of adding two or more numbers.

table—Information arranged in columns and rows so that is can be easily understood.

take-home pay—The amount of pay which remains after deductions have been made.

tax—An amount of money charged by the government on income, property or services.

total—The sum or product of a list of amounts.

unit cost—The actual amount which has been paid by the seller for an item which he will sell; such amounts are usually marked up for profit.

unit price—The amount that the buyer has to pay for one item.

value—What something is worth.

watt—Unit for measuring electric power.

withdraw—To take money out of a bank account.

withdrawal—An amount which is withdrawn.